Vibrations of Linear Piezostructures

Wiley-ASME Press Series

Computer Vision for Structural Dynamics and Health Monitoring
Dongming Feng, Maria Q Feng

Theory of Solid-Propellant Nonsteady Combustion
Vasily B. Novozhilov, Boris V. Novozhilov

Introduction to Plastics Engineering
Vijay K. Stokes

Fundamentals of Heat Engines: Reciprocating and Gas Turbine Internal Combustion Engines
Jamil Ghojel

Offshore Compliant Platforms: Analysis, Design, and Experimental Studies
Srinivasan Chandrasekaran, R. Nagavinothini

Computer Aided Design and Manufacturing
Zhuming Bi, Xiaoqin Wang

Pumps and Compressors
Marc Borremans

Corrosion and Materials in Hydrocarbon Production: A Compendium of Operational and Engineering Aspects
Bijan Kermani and Don Harrop

Design and Analysis of Centrifugal Compressors
Rene Van den Braembussche

Case Studies in Fluid Mechanics with Sensitivities to Governing Variables
M. Kemal Atesmen

The Monte Carlo Ray-Trace Method in Radiation Heat Transfer and Applied Optics
J. Robert Mahan

Dynamics of Particles and Rigid Bodies: A Self-Learning Approach
Mohammed F. Daqaq

Primer on Engineering Standards, Expanded Textbook Edition
Maan H. Jawad and Owen R. Greulich

Engineering Optimization: Applications, Methods and Analysis
R. Russell Rhinehart

Compact Heat Exchangers: Analysis, Design and Optimization using FEM and CFD Approach
C. Ranganayakulu and Kankanhalli N. Seetharamu

Robust Adaptive Control for Fractional-Order Systems with Disturbance and Saturation
Mou Chen, Shuyi Shao, and Peng Shi

Robot Manipulator Redundancy Resolution
Yunong Zhang and Long Jin

Stress in ASME Pressure Vessels, Boilers, and Nuclear Components
Maan H. Jawad

Combined Cooling, Heating, and Power Systems: Modeling, Optimization, and Operation
Yang Shi, Mingxi Liu, and Fang Fang

Applications of Mathematical Heat Transfer and Fluid Flow Models in Engineering and Medicine
Abram S. Dorfman

Bioprocessing Piping and Equipment Design: A Companion Guide for the ASME BPE Standard
William M. (Bill) Huitt

Nonlinear Regression Modeling for Engineering Applications: Modeling, Model Validation, and Enabling Design of Experiments
R. Russell Rhinehart

Geothermal Heat Pump and Heat Engine Systems: Theory and Practice
Andrew D. Chiasson

Fundamentals of Mechanical Vibrations
Liang-Wu Cai

Introduction to Dynamics and Control in Mechanical Engineering Systems
Cho W. S. To

Vibrations of Linear Piezostructures

Andrew J. Kurdila and Pablo A. Tarazaga
Virginia Polytechnic Institute and State University

This Work is a co-publication between John Wiley & Sons Ltd
and ASME Press

WILEY

ASME PRESS

This edition first published 2021

This Work is a co-publication between John Wiley & Sons Ltd and ASME Press

Registered Office
John Wiley & Sons, Inc., 111 River Street, Hoboken, NJ 07030, USA

Editorial Office
111 River Street, Hoboken, NJ 07030, USA

For details of our global editorial offices, customer services, and more information about Wiley products visit us at www.wiley.com.

Wiley also publishes its books in a variety of electronic formats and by print-on-demand. Some content that appears in standard print versions of this book may not be available in other formats.

Library of Congress Cataloging-in-Publication Data

Names: Kurdila, Andrew, author. | Tarazaga, Pablo (Pablo A.), author.
Title: Vibrations of linear piezostructures / Andrew J. Kurdila and Pablo A. Tarazaga.
Description: First edition. | Hoboken, NJ : John Wiley & Sons, 2021. | Series: Wiley-ASME Press series | Includes bibliographical references and index.
Identifiers: LCCN 2020027699 (print) | LCCN 2020027700 (ebook) | ISBN 9781119393405 (cloth) | ISBN 9781119393504 (adobe pdf) | ISBN 9781119393528 (epub) | ISBN 9781119393382 (obook)
Subjects: LCSH: Piezoelectricity. | Vibration.
Classification: LCC QC595 .K78 2021 (print) | LCC QC595 (ebook) | DDC 537/.2446–dc23
LC record available at https://lccn.loc.gov/2020027699
LC ebook record available at https://lccn.loc.gov/2020027700

Cover Design: Wiley
Cover Image: Pablo A. Tarazaga

Set in 9.5/12.5pt STIXTwoText by SPi Global, Chennai, India
Printed and bound by CPI Group (UK) Ltd, Croydon, CR0 4YY

C9781119393405_260421

Contents

Foreword

The rise of piezoelectric materials as sensors and actuators in engineering systems got started around 1980 and began to make an impact in the world of vibrations about five years after that. Subsequently, it started to explode into the 90s with topics such shunt damping, active control, structural health monitoring and energy harvesting. As a result, the need to document the fundamentals and intricacies of modeling piezoelectric materials in the context of vibrations in book form will well serve a variation of communities. The presentation here puts the topic on a firm mathematical footing.

The authors are uniquely qualified to provide a sophisticated analytical framework with an eye for applications. Professor Kurdila has nearly four decades of experience in modeling of multi-physics systems. He authored two other books, one on structural dynamics, and several research monographs. Professor Tarazaga is an experienced creator of piezoelectric solutions to vibration and control problems. Both are well published in their respective research areas of research. Their combined expertise in researching vibratory systems integrated with piezoelectric materials enables this complete and detailed book on the topic. This allows for a formal theoretical background which will enable future research.

Daniel J. Inman
Ann Arbor, Michigan

Preface

The goal of this book is to provide a self-contained, comprehensive, and introductory account of the modern theory of vibrations of linearly piezoelectric structural systems. While the piezoelectric effect was first investigated by the Curies in the 1880s, and systematically investigated in the field of acoustics and the development of sonar during the First World War, it is only much more recently that we have seen the widespread interest in mechatronic systems that feature piezoelectric sensors and actuators. Many of the early, now classical, texts present piezoelectricity from the viewpoint of a material scientist such as in [22] or [53]. Others are difficult, if not impossible, to obtain since they are out of print. Older editions of the excellent text [20] are currently selling for prices in excess of $600 on sites such as Amazon.com. Moreover, it is also quite difficult to find treatments of piezoelectricity that systematically cover all the relevant background material from first principles in continuum mechanics, continuum electrodynamics, or variational calculus that are necessary for a comprehensive introduction to vibrations of piezoelectric structures. The authors know of no text that assimilates all this requisite supporting material into one source. One text may give an excellent overview of piezoelectric constitutive laws, but neglect to discuss variational methods. Another may cover variational methods for piezoelectric systems, but fail to review the first principles of electrodynamics, and so forth. A large, substantive literature on various aspects of piezoelectricity has evolved over the past few years in archival journal articles, but much of this material has never been systematically represented in a single text.

This book has evolved from the course notes that the authors have generated while offering courses in active materials, smart systems, and piezoelectric materials over the past decade at various research universities. Most recently, the authors have taught active materials and smart structures courses that feature piezoelectricity at Virginia Tech to a diverse collection of first year graduate students. So much time was dedicated to the particular systems that include piezoelectric components that this textbook emerged. The backgrounds of the students in our classes have varied dramatically. Many students have not had a graduate class in vibrations, continuum mechanics, advanced strength of materials, nor electrodynamics. For this reason, the notes that evolved into this book make every effort to be self-contained. Admittedly, this text covers in one chapter what other courses may cover over one or two semesters of dedicated study. As an example, Chapter 3 reviews the fundamentals of continuum mechanics for this text, a topic that is covered in other graduate classes at an introductory level during a full semester. So, while the presentation attempts to be comprehensive, the pace is sometimes brisk.

While preparing this text, we have tried to structure the material so that it is presented at the senior undergraduate or first year graduate student level. It is intended that this text provide the student with a good introduction to the topic, one that will serve them well when they seek to pursue more advanced topics in other texts or in their research. For example, this text can serve as a introduction to the fundamentals of modeling piezoelectric systems, and it can prepare the student specializing in energy harvesting when they consult a more advanced text such as [21].

This text begins in Chapter 2 with a review of the essential mathematical tools that are used frequently throughout the book. Topics covered include frames, coordinate systems, bases, vectors, tensors, introductory crystallography, and symmetry. Chapter 3 then gives a fundamental summary of topics from continuum mechanics. The stress vector and tensor is defined, Cauchy's Principle and the equilibrium equations are derived. The strain tensor is defined, and an introduction to constitutive laws for linearly elastic materials is also covered in this chapter. Chapter 4 provides the student the required introduction to continuum electrodynamics that is essential in building the theory of linear piezoelectricity in subsequent chapters. The definitions of charge, current, electric field, electric displacement, and magnetic field are introduced, and then Maxwell's equations of electromagnetism are studied.

Linear piezoelectricity is covered in Chapter 5. The discussion begins by introducing a physical example of the piezoelectric effect in one spatial example, and subsequently giving a generalization of the phenomenon in terms of piezoelectric constitutive laws. The initial-boundary value problem of linear piezoelectricity is then derived from the analysis of Maxwell's equations and principles of continuum mechanics. While the equations governing any particular piezoelectric structure can be derived in principle from the initial-boundary value problem of linear piezoelectricity, it is often possible and convenient to derive them directly for a problem at hand. Chapter 6 discusses the application of Newton's equations of motion for several prototypical piezoelectric composite structural systems. Chapter 7 provides a detailed account of how variational techniques can be used, instead of Newton's method, for many linearly piezoelectric structures. In some cases the variational approach can be much more expedient in deriving the governing equations. This chapter starts with a review of variational methods and Hamilton's Principle for linearly elastic structures. The approach is then extended by formulating Hamilton's Principle for Piezoelectric Systems and Hamilton's Principle for Electromechanical Systems. Several examples are considered, including the piezoelectrically actuated rod and Bernoulli–Euler beam, as well as the electromechanical systems that result when these structures are connected to ideal passive electrical networks. The book finishes in Chapter 8 with a discussion of approximation methods. Both modal approximations and finite element methods are discussed. Numerous example simulations are described in the final chapter, both for the actuator equation alone and for systems that couple the actuator and sensor equations.

June, 2017

Andrew J. Kurdila
Pablo A. Tarazaga

Acknowledgments

This book is the culmination of research carried out and courses taught by the authors over the years at a variety of institutions. The authors would like to thank the various research laboratories and sponsors that have supported their efforts over the years in areas related to active materials, smart structures, linearly piezoelectric systems, vibrations, control theory, and structural dynamics. These sponsors most notably include the Army Research Office, Air Force Office of Scientific Research, Office of Naval Research, and the National Science Foundation. We likewise extend our appreciation to the institutes of higher learning that have enabled and supported our efforts in teaching, research, and in disseminating the fruits of teaching and research: this volume would not have been possible without the infrastructure that makes such a sustained effort possible. In particular, we extend our gratitude to the Aerospace Engineering Department at Texas A&M University, the Department of Mechanical and Aerospace Engineering at the University of Florida, and most importantly, the Department of Mechanical Engineering at Virginia Tech. We extend our appreciation to the many colleagues that have worked with us over the years in areas related to active materials and smart structures. In particular, we thank Dr. Dan Inman for his support and for being a source of inspiration.

We also would like to specifically thank Dr. Vijaya V. N. Sriram Malladi and Dr. Sai Tej Paruchuri for their tireless efforts in editing and correcting the draft manuscript. Their meticulous attention to detail, suggestions and tireless effort has made this book a better version from its original draft. Additionally, we would like to thank our students Dr. Sheyda Davaria, Dr. Mohammad Albakri, Manu Krishnan, Mostafa Motaharibidgoli who have worked through the manuscript in order to improve its clarity. We would also like to also thank Sourabh Sangle, Murat Ambarkutuk, Lucas Tarazaga and Vanessa Tarazaga for their help in proofreading the last draft of the document. Finally, we would like to acknowledge anyone else not mentioned that contributed to the manuscript, including the students in our classes who provided valuable input throughout the years.

And, of course, we thank our families for their continued support and encouragement in efforts just like this one over the years.

Andrew J. Kurdila
Pablo A. Tarazaga

Blacksburg, VA
February, 2021

List of Symbols

Symbol	Description
Vectors and Tensors	
δ_{ij}	Kronecker delta function
ϵ_{ijk}	Levi-Civita permutation tensor
$\boldsymbol{g}_i$	generic basis vector
$\boldsymbol{R}, r_{ij}$	rotation matrix and its components
$V \otimes V \otimes \cdots \otimes V$	vector space of nth order tensors
$\boldsymbol{g}_i \otimes \boldsymbol{g}_j$	tensor product of $\boldsymbol{g}_i$ and $\boldsymbol{g}_j$
χ_Ω	characteristic function of Ω
$\boldsymbol{a}, \boldsymbol{b}, \boldsymbol{c}$	lattice parameters
α, β, γ	unit cell or lattice angles
Ω	domain
$\partial\Omega$	boundary of Ω
Electrodynamics	
c	speed of light
ϵ_0	electric permitivity of free space
μ_0	magnetic permeability of free space
i	current
$\boldsymbol{j}, \boldsymbol{j}_f, \boldsymbol{j}_b, \boldsymbol{j}_p$	total, free, bound, and polarization current density
$\rho_e, \rho_{ef}, \rho_{e,b}$	total, free, and bound charge density
ϕ	electric potential
$\boldsymbol{\Psi}$	magnetic vector potential
$\boldsymbol{E}, E_i$	electric field vector and its components
$\boldsymbol{p}, p_i$	dipole moment and its components
$\boldsymbol{P}, P_i$	polarization and its components

Symbol	Description
$\boldsymbol{D}, D_i$	electric displacement vector and its components
$\boldsymbol{B}, B_i$	magnetic field and its components
$\boldsymbol{m}, m_i$	magnetic dipole moment and its components
$\boldsymbol{M}, M_i$	magnetization or magnetic polarization
$\boldsymbol{H}, H_i$	magnetic field intensity and its components
$\boldsymbol{E}_e, E_{e,j}$	"external" electric field induced by free charge
$\boldsymbol{E}_i, E_{i,j}$	"internal" electric field induced by polarization charge
Elasticity	
ρ_m	mass density
$\boldsymbol{f}_b, f_{bi}$	body force and its components
$\boldsymbol{u}, u_i$	displacement field and its components
$\boldsymbol{T}, T_{ij}$	second order stress tensor and its components
$\boldsymbol{S}, S_{ij}$	second order linear strain tensor and its components
$\boldsymbol{C}, C_{ijkl}$	fourth order material stiffness tensor and its components
$\partial\Omega_u$	boundary of Ω on which u_i is prescribed
$\partial\Omega_T$	boundary of Ω on which T_{ij} is prescribed
$\overline{u}_i$	prescribed displacements on $\partial\Omega_u$
$\overline{\tau}_i$	prescribed stress vector on $\partial\Omega_T$
$\overline{u}_{i,0}$	initial condition on u_i in Ω
$\overline{v}_{i,0}$	initial condition on $\frac{\partial u_i}{\partial t}$ in Ω
$\mathcal{U}_0$	strain energy density
$\mathcal{V}$	strain energy or potential energy
$\mathcal{T}$	Kinetic energy
W	Work
δW	Virtual work
$\mathcal{S}, \mathcal{M}$	Beam shear force and bending moment
$\mathcal{M}_i$	Plate bending moment per unit length
$\mathcal{S}_i$	Plate shear force per unit length
κ	beam area moment
$C_{11}^E I$	Beam bending stiffness
Piezoelectricity	
$\partial\Omega_\phi$	boundary of Ω on which ϕ is prescribed
$\partial\Omega_D$	boundary of Ω on which D_i is prescribed
$\overline{\phi}$	prescribed potential ϕ on $\partial\Omega_\phi$

Symbol	Description
$\overline{\sigma}$	prescribed charge distribution on $\partial\Omega_D$
$\mathcal{Q}$	heat
$\mathcal{U}$	internal energy
$\mathcal{E}$	total electromechanical energy
$\mathcal{H}$	electric enthalpy density
$\mathcal{V}_{\mathcal{H}}$	electric enthalpy
V	voltage
Q	abstract (vector) space of generalized coordinates
$\mathcal{F}$	time varying trajectories that take value in Q
C^E_{ijkl}	4th order tensor, constant $\boldsymbol{E}$, Eq. 5.19
d_{nij}	3rd order piezoelectric tensor in Eq. 5.19
ϵ^S_{ij}	2nd order tensor, constant $\boldsymbol{S}$, in Eq. 5.19
C^D_{ijkl}	4th order tensor, constant $\boldsymbol{D}$, Eq. 5.14
e_{nij}	3rd order piezoelectric tensor in Eq. 5.14
β^S_{ij}	2nd order tensor, constant $\boldsymbol{S}$, in Eq. 5.14
Θ	absolute temperature
s	entropy
$J(q)$	functional $J : \mathcal{F} \rightarrow \mathbb{R}$
$\mathcal{A}(q)$	Action integral of Hamilton's Principle
$\mathcal{A}_{\mathcal{H}}(q)$	Action integral, Hamilton's Principle of Piezoelectricity
$DJ(q, p)$	Gateaux derivative at $p \in Q$ in the direction $q \in Q$
$\delta(\cdot)$	(virtual) Variation operator
$\mathcal{V}_{em}$	Electromechanical potential
$\mathcal{V}_e$	Electrical potential of ideal capacitors
$\mathcal{V}_m$	Magnetic potential of ideal inductors
$\mathfrak{L}$	Lagrangian density

1
Introduction

1.1 The Piezoelectric Effect

In the most general terms, a material is piezoelectric if it transforms electrical into mechanical energy, and vice versa, in a reversible or lossless process. This transformation is evident at a macroscopic scale in what are commonly known as the direct and converse piezoelectric effects. The direct piezoelectric effect refers to the ability of a material to transform mechanical deformations into electrical charge. Equivalently, application of mechanical stress to a piezoelectric specimen induces flow of electricity in the direct piezoelectric effect. The converse piezoelectric effect describes the process by which the application of an electrical potential difference across a specimen results in its deformation. The converse effect can also be viewed as how the application of an external electric field induces mechanical stress in the specimen.

While the brothers Pierre and Jacques Curie discovered piezoelectricity in 1880, much the early impetus motivating its study can be attributed to the demands for submarine countermeasures that evolved during World War I. An excellent and concise history, before, during, and after World War I, can be found in [43]. With the increasing military interest in detecting submarines by their acoustic signatures during World War I, early research often studied naval applications, and specifically sonar. Paul Langevin and Walter Cady had pivotal roles during these early years. Langevin constructed ultrasonic transducers with quartz and steel composites. Shortly thereafter, the use of piezoelectric quartz oscillators became prevalent in ultrasound applications and broadcasting. The research by W.G. Cady was crucial in determining how to employ quartz resonators to stabilize high frequency electrical circuits.

Vibrations of Linear Piezostructures, First Edition. Andrew J. Kurdila and Pablo A. Tarazaga.

This Work is a co-publication between John Wiley & Sons Ltd and ASME Press.

A number of naturally occurring crystalline materials including Rochelle salt, quartz, topaz, tourmaline, and cane sugar exhibit piezoelectric effects. These materials were studied methodically in the early investigations of piezoelectricity. Following World War II, with its high demand for quartz plates, research and development of techniques to synthesize piezoelectric crystalline materials flourished. These efforts have resulted in a wide variety of synthetic piezoelectrics, and materials science research into specialized piezoelectrics continues to this day.

1.1.1 Ferroelectric Piezoelectrics

Perhaps one of the most important classes of piezoelectric materials that have become popular over the past few decades are the ferroelectric dielectrics. A ferroelectric can have coupling between the mechanical and electrical response that is several times a large as that in natural piezoelectrics. Ferroelectrics include materials such as barium titanate and lead zirconate titanate, and their unit cells are depicted in Figure 1.1. When the centers of positive and negative charge in a unit cell of a crystalline material do not coincide, the material is said to be polar or dielectric. An electric dipole moment $\boldsymbol{p}$ is a vector that points from the center of negative charge to the center of positive charge, and its magnitude is equal to $|\boldsymbol{p}| = q \cdot \delta$ where q is the magnitude of the charge at the centers and δ is the separation between the centers. The limiting volumetric density of dipole moments is the polarization vector $\boldsymbol{P}$. Intuitively we think of the polarization vector $\boldsymbol{P}$ as measuring the asymmetry of the internal electric field of the piezoelectric crystal lattice. Ferroelectrics exhibit spontaneous electric polarization that can be reversed by the

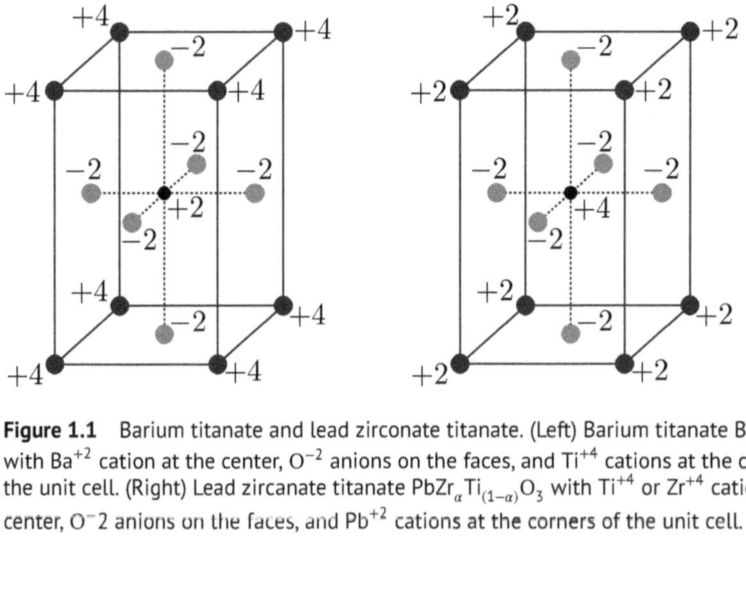

Figure 1.1 Barium titanate and lead zirconate titanate. (Left) Barium titanate $BaTiO_3$ with Ba^{+2} cation at the center, O^{-2} anions on the faces, and Ti^{+4} cations at the corners of the unit cell. (Right) Lead zircanate titanate $PbZr_{\alpha}Ti_{(1-\alpha)}O_3$ with Ti^{+4} or Zr^{+4} cation at the center, $O^{-}2$ anions on the faces, and Pb^{+2} cations at the corners of the unit cell.

application of an external electric field. In other words, the polarization of the material is evident during a spontaneous process, one that evolves to a state that is thermodynamically more stable. Understanding this process requires a discussion of the micromechanics of a ferroelectric.

The micromechanics of ferroelectric dielectrics is subtle and interesting. Above a critical temperature T_c, the Curie temperature, the crystal structure of a ferroelectric is usually symmetric, and a plot of the polarization versus applied electric charge is generally nonlinear and single-valued as shown in Figure 1.2.

However, with cooling below the Curie temperature T_c, a thermodynamic process drives a structural phase transition so that the final crystalline phase has a lower symmetry. At the lower temperature it can be shown [18] that the lower symmetry crystal phase has at least two energetically equivalent configurations or variants. Furthermore, with the application of an external electric field, it must be the case that it is possible switch among these crystalline variants in a reversible process. The ferroelectric material forms *domains* that consist of these energetically equivalent crystalline variants. Figure 1.3 depicts schematically the 180° and 180°/90° domains [31] that can appear in single crystal barium titanate $BaTiO_3$ [31]. Note in the figure that the polarization vectors are opposite from one domain to the next, and their average polarization over a macroscale can have zero effective polarization. Because of the presence of these domains, below the Curie

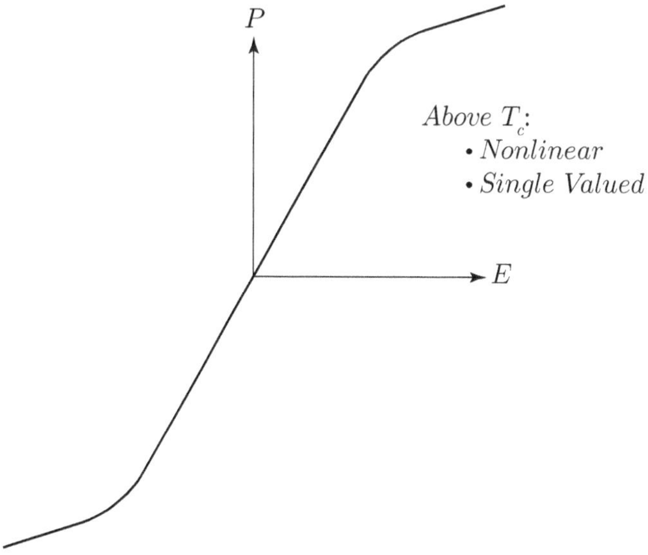

Figure 1.2 Polarization versus applied electrical field for ferroelectric above the Curie temperature T_c.

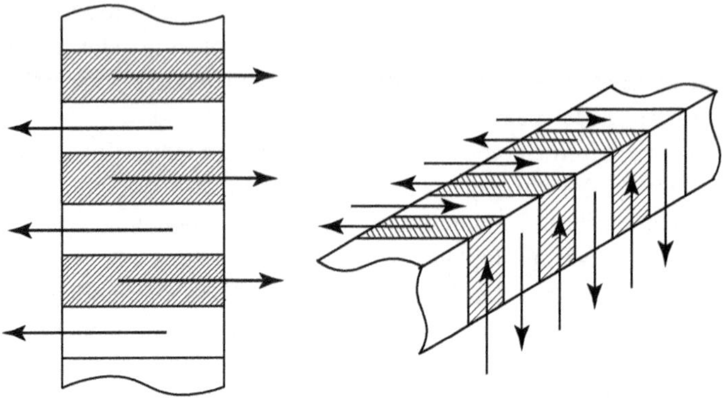

Figure 1.3 180° and 180°/90° domains in $BaTiO_3$, [31]. Source: Walter J. Merz, Domain Formation and Domain Wall Motion in Ferro-electric $BaTiO_3$ Single Crystals, em Physical Review, Volume 95, Number 3, August 1, 1954, pp. 690–698.

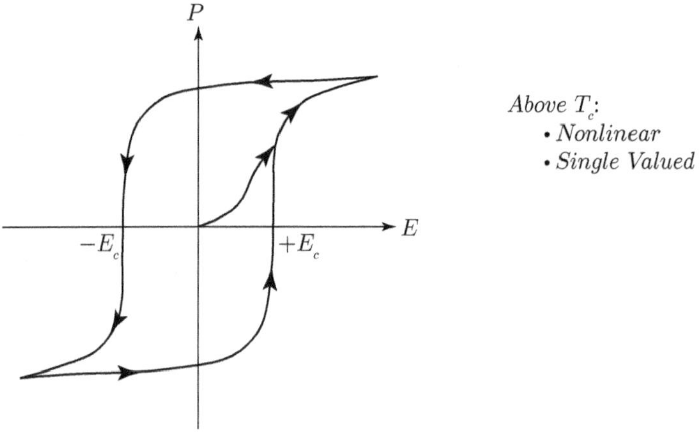

Figure 1.4 Polarization versus electrical field hysteresis below the Curie temperature T_c.

temperature T_c the polarization versus applied electric field takes the form of a hysteresis loop as shown in Figure 1.4. Initially, the domains cancel their effects over the macroscopic specimen and $P = 0$ at $E = 0$. The polarization P increases as in Figure 1.2 for a range of electric field E. When a critical value E_c, the coercive electric field strength, is reached, the domains abruptly switch so that they are approximately well-aligned with the external electric field. With all domains having approximately aligned polarization vectors, the polarization again follows a nonlinear single valued curve until saturation is achieved. When the electric field is reversed, and reaches the opposite coercive electric field strength $-E_c$, the domains switch again so their polarization vectors are approximately aligned with the second variant. The result of this cyclic process is that after the transient response there is a nonzero polarization, the spontaneous polarization, for an electric field strength $E = 0$. At a macroscopic scale, then, the effective or average polarization can switch with the application of the external electric field.

1.1.2 One Dimensional Direct and Converse Piezoelectric Effect

In view of these observations, at a fundamental level, the micromechanics of piezoelectricity is understood in terms of crystalline asymmetry. While the most general theory of linear piezoelectricity of material continua in three dimensions is discussed in Chapter 5, intuition can be built by considering a one dimensional example. Figure 1.5 depicts the direct piezoelectric effect graphically, while the converse effect is shown in Figure 1.6. For the specimens shown, the mechanical variables are the stress T and strain S, and the electrical variables include the electric field E, electric displacement D, voltage $V = \phi^+ - \phi^-$, and the electrical potential ϕ. In Figure 1.5 we suppose that the top and bottom of the specimen are free to displace. A thin film electrode, one that does not alter the mechanical properties of the specimen, is applied to the top and bottom surfaces by a deposition or sputtering process. An ideal current meter, over which the potential difference is approximately zero, is attached to the top and bottom electroded surfaces. A positive stress T is applied as shown. As we discuss in Chapter 5 the constitutive laws that couple the electrical and mechanical variables can take many forms. In this one dimensional example we choose to express the dependency among the electrical and mechanical variables as

$$\begin{Bmatrix} S \\ D \end{Bmatrix} = \begin{bmatrix} s & d \\ d & \varepsilon \end{bmatrix} \begin{Bmatrix} T \\ E \end{Bmatrix}. \tag{1.1}$$

where s is the mechanical compliance constant, d is the piezoelectric coupling constant, and ε is the permittivity constant. See Chapter 5 for a precise definition of these constants when interpreted as elements of tensors suitable for piezoelectric continua in three dimensions. Since the current meter is ideal, the electrical potential ϕ at the top electrode is equal to the potential at the bottom electrode, and the electric field $\boldsymbol{E} = -\nabla\phi \approx -(\phi^+ - \phi^-)/t \approx 0$. From Equation 1.1 it follows that the strain $S = sT$ is positive, and the top and bottoms of the specimens displace by approximately $u \approx S \cdot t/2$, and the specimen stretches by a total amount $\approx S \cdot t$. The magnitude of the charge on the electrodes can be obtained from the boundary conditions $|q| = DA = dTA$. Thus, we see that either the application of a positive stress T, or equivalently of a positive strain $S = sT$, induces charges having magnitude $|q| = dTA$ on the faces of the specimen.

In contrast, Figure 1.6 illustrates how the application of an electric potential across a piezoelectric specimen generates a deformation. In this case, we assume

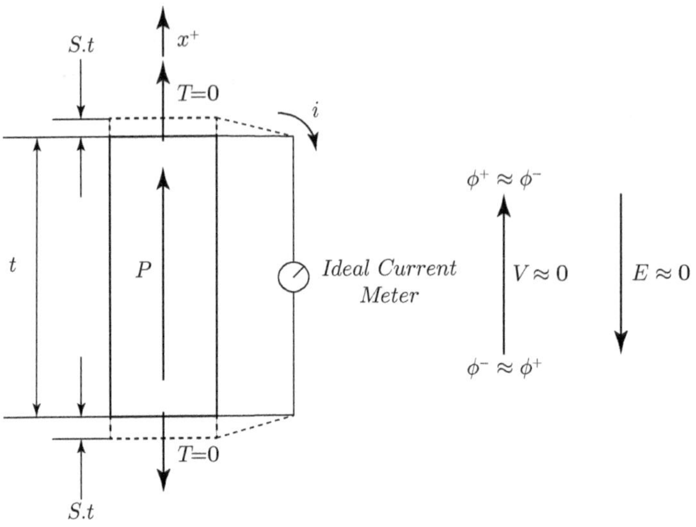

Figure 1.5 The direct piezoelectric effect.

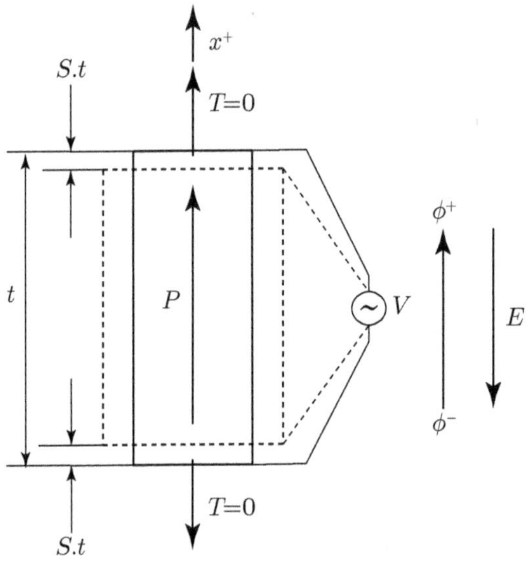

Figure 1.6 The converse piezoelectric effect.

that the potential at the top electrode is ϕ^+, the potential at the lower electrode is ϕ^-, and the voltage source has a voltage $V = \phi^+ - \phi^-$. Since the stress $T = 0$ and $\boldsymbol{E} = -\nabla\phi$, the electrical field E is approximately given by $E = -V/t$, and the strain in the specimen at its top and bottom is $S = dE = -dV/2t$. In this case, the specimen compresses in total approximately by $-dV$.

1.2 Applications

While the earliest applications of the piezoelectric effect were mostly related to ultrasound generation or sound measurements, the commercially viable applications of piezoelectricity have grown remarkably in diversity since that time. A quick survey of companies such as PCB® or PI Ceramics® that offer a broad range of piezoelectric devices can quickly convey the breadth of applications. For example, PI Ceramics notes that a piezoelectric device finds use in "*...high-end technology markets, such as medical technology, mechanical and automotive engineering or semiconductor technology, but is also present in everyday life, for example as a generator of ultrasonic vibrations in a cleaning bath for glasses and jewelry or in medical tooth cleaning.*" The companies PCB® [36] and PI Ceramics® [37] include descriptions of broad applications areas. Even within each of these areas, the specific inventory of applications can be vast. Accelerometers, for example, are but one specific class of piezoelectrically-based sensor that are used to measure vibration, shock, and acceleration for testing, control, online estimation, and system monitoring. The company PCB® [36] notes on `http://www.pcb.com/TestMeasurement/Accelerometers` that their products in this niche of sensors support applications in "*...balancing, bearing, analysis, biomechanics, building vibration monitoring, biomechanics, bridge monitoring, component durability testing, crash testing, drop testing, fatigue testing, gearbox monitoring, environmental testing, ground vibration testing, impact measurements, impulse response measurements, machinery vibration, modal analysis, package testing, product qualification, quality control, seismic monitoring, structural testing, structure-borne noise, vibration analysis, vibration isolation, and vibration stress screening.*"

Despite the growth of the applications of piezoelectric technology catalogued above, it is common that they are often grouped into three general categories: energy applications, sensors, and actuators or motors. We discuss these next.

1.2.1 Energy Applications

While small piezoelectric specimens typically to do not output large currents or power, they can generate very high voltage differences under application of stress. It is for this reason that piezoelectric igniters are commonly used in gas broilers, gas stoves and ranges, gas fireplaces, or other appliances. The igniters usually operate by releasing a spring loaded switch that impacts the piezoelectric specimen, thereby generating a large, transient potential difference [36]. Such a voltage is high enough that it induces current flow across a gap that ignites the gas. In addition, the use of piezoelectric transducers in energy harvesting applications is a rapidly growing field. See for example [14] and the references therein for a good, comprehensive technical treatment of this topic. As noted before, while small piezoelectric transducers *usually* are not appropriate as large supplies of power, they are well suited to applications that require local, modular, or isolated energy sources for microscale electromechanical (MEMs) devices. Often, classes of sensors require small sources of energy to perform their associated measurements, and piezoelectric transducers have proven to be a viable route to support such sensors. There are numerous studies of energy reclamation from a wide range of sources [14]. Examples of composite piezoelectric structures that are connected to linear, ideal, passive electrical networks are studied in Chapters 6, 7, and 8. These electromechanical models are also suitable for the study of nonlinear switching strategies for energy harvesting. The emerging field of MEMs or NEMs (nanoelectromechanical systems) robotics requires microscale energy supplies to enable their mobility, and piezoelectric transducers are often the choice to develop self-contained MEMs robots.

1.2.2 Sensors

Among all the various categories of applications of piezoelectric devices and systems, it is perhaps the piezoelectric sensors that have had the most profound impact on measurement technology infrastructure. As noted above, there are a wide and growing collection of piezoelectric transducers that serve as sensors. These include pressure transducers, load cells or force transducers, torque transducers, acoustic microphones, accelerometers, and vibration sensors. The underlying physical basis of nearly all these sensors is the same: the forces applied on opposite sides of a piezoelectric specimen generate a source of charge that is highly sensitive to the applied force, and the measurement of the charge or current provides an estimate of such forces over extremely high frequencies. Often the bandwidth of such sensors extends to the hundreds of kilohertz range. The diversity of piezoelectrically-based, commercial microphones available from PCB® is illustrated in Figure 1.7. This figure illustrates that the operating range of these sensors is as high as O(100) kHz, and their sensitivity varies from O(1) mV/Pa to O(10) mV/Pa. The physics underlying a wide range of piezoelectric sensors is studied in Chapters 3, 4, 5, 6, 7, and 8.

1.2.3 Actuators or Motors

Piezoelectric actuators or motors are available commercially at companies such as PCB® or PI Ceramics®, and they have been the topic of research in academia and national laboratories for decades. The most popular commercially available actuators are the linear or stack actuators and the bender actuators. Stack actuators

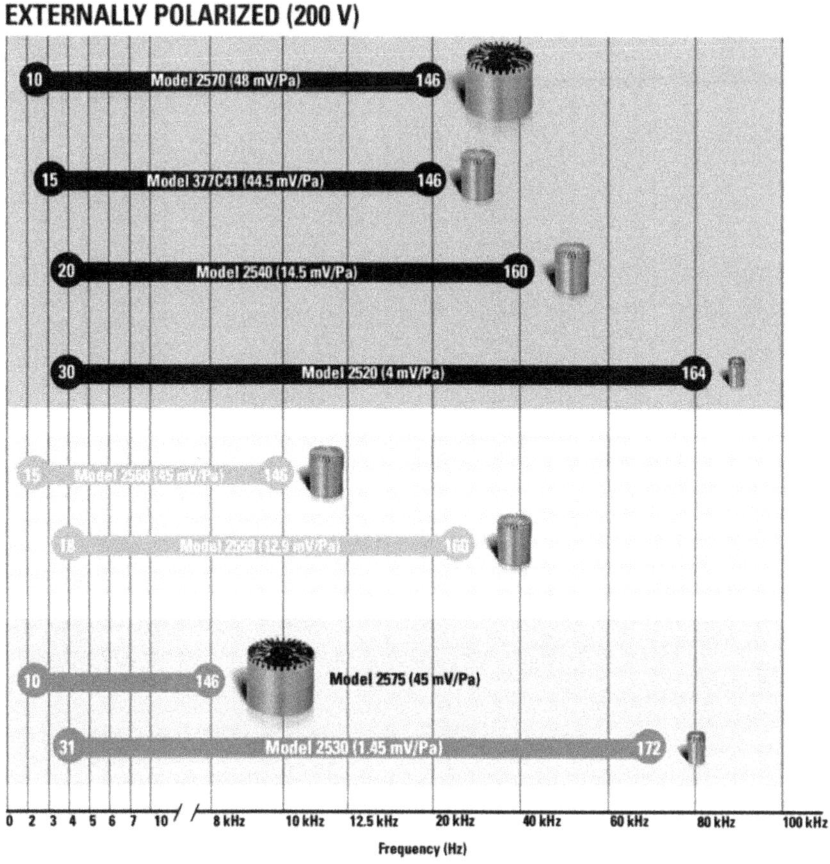

Figure 1.7 Piezoelectrically based microphones from PCB®, Source: HOW TO SELECT THE BEST TEST & MEASUREMENT PRECISION MICROPHONE, PCB Piezotronics, Inc. Retrieved from: http://www.pcb.com/Microphones Preamplifiers Acoustic Accessories/select.

developed by PI Ceramic® are depicted in Figure 1.8. Commercially available stack actuators usually have a small stroke, on the order O(10) to O(100) μm. These prototypical piezoelectric composites are studied in Chapters 5, 6, 7, and 8. Figure 1.9 depicts bender actuators available from PI Ceramics®. The bending actuators are intended for applications that require a larger stroke and operate at a lower frequency than the linear actuators. The stroke of the benders depicted in Figure 1.9 varies approximately from O(100) to O(1000) μm. Composite bending actuators are studied in Chapters 5, 6, 7, and 8.

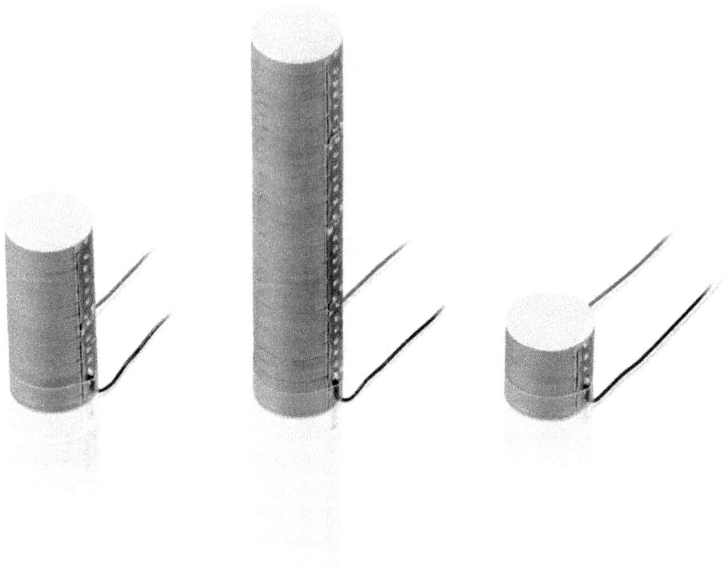

Figure 1.8 Piezoelectric stack actuators available from PI ceramic®, Source: GmbH & Co. KG owner.

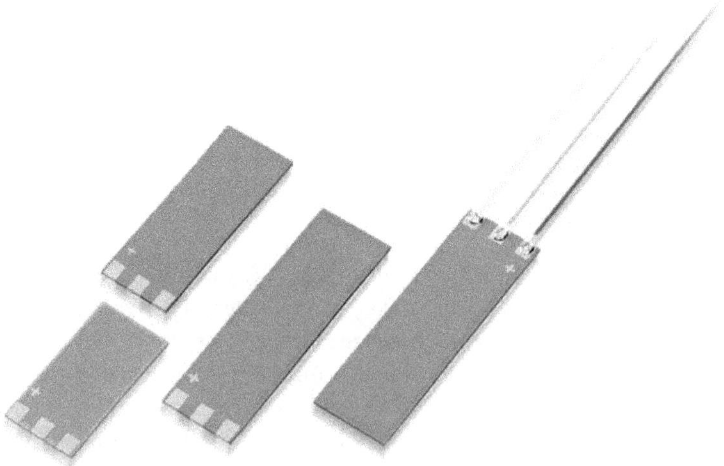

Figure 1.9 Piezoelectric bender actuators available from PI ceramic®, Source: GmbH & Co. KG owner.

1.3 Outline of the Book

The linear and bending actuators shown in Figure 1.8 and 1.9 are considered in examples throughout this text: they are perhaps the most simple macroscale actuators that are commercially available. In fact, they can also be used to understand the basic physics of piezoelectric sensors too. They illustrate the operating principles of and modeling techniques for piezoelectric composites that are evident in much more complex designs.

Unfortunately, the theoretical underpinnings of piezoelectric mechanics embraces a wide collection of fields of study that must be synthesized. We begin in Chapter 2. Section 2.1 reviews the fundamentals of vectors, bases, and frames of reference. This section is vital in developing an understanding of how physical vectors such as velocity, acceleration, stress vectors, electric field vectors, and electric displacement vectors are represented. The section culminates in a presentation of rotation matrices and their essential role in constructing change of bases for different representations of vectors. Section 2.2 then extends these results by introducing multilinear operators, or tensors, that act between vector spaces. Since vectors are first order tensors, they are a special case of the tensors presented in Section 2.2. In addition, another collection of physical variables critical to linear piezoelectricity are understood as tensors. These include the stress tensor, linear strain tensor, permittivity tensor, piezoelectric coupling tensor, stiffness tensor, and compliance tensor. The section concludes with a discussion of the role of rotation matrices in the representation and change of basis formula for n^{th} order tensors. Section 2.3 discusses symmetry properties and geometric properties of tensors and crystals. The discussion begins with an overview of the geometry of crystals in Section 2.3.1. The 14 Bravais lattices and seven crystal systems are defined, as are the 32 crystallographic point groups. This section concludes with examples of symmetry transformations for typical crystal classes, and a discussion of tensor invariance associated with symmetry operations.

Chapter 3 reviews the basics of continuum mechanics that are needed to build a coherent framework for linear piezoelectricity. The definition of the stress tensor in three dimensions is presented in Section 3.1.1, Cauchy's formula is presented in Section 3.1.2, and the equations of equilibrium are discussed in Section 3.1.3. Section 3.2 is dedicated to the study of the linear strain tensor, the mechanical field variable that is complementary to the stress tensor. The general definition of the linear strain tensor in three dimensions is presented, as well as the kinematic relationships for common structures such as the axial rod, Bernoulli–Euler beam, and Kirchoff plate. Strain energy is discussed in Section 3.3. The strain energy density function is introduced, and its role in determining additional symmetry properties of the material stiffness tensor is given. A review of the constitutive laws for linearly elastic materials is summarized in Section 3.4, and the structure of the constitutive relationships for triclinic, orthotropic, and transversely isotropic materials are summarized. The chapter ends with an introduction of the initial-boundary value problem of linear elasticity in Section 3.5.

We turn to a discussion of continuum electrodynamics in Chapter 4. Charge and current are introduced in Section 4.1, and the static electric and magnetic fields are discussed in Section 4.2. Maxwell's equations are introduced in Equation 4.10 in SI units. Section 4.3.1 relates the polarization and electric displacement vectors, and relates them to bound and mobile charge, respectively. Magnetization and magnetic field intensity are defined in Section 4.3.2, as are the free, bound, and polarization current densities, respectively. Section 4.3.3 discusses the form of Maxwell's equations in Gaussian units, which prove to be convenient for the derivation of the equations of piezoelectricity.

Chapter 5 presents the theory of linear piezoelectricity, starting with some one dimensional examples in Section 5.1. Section 5.2 gives the detailed account of how the equations of linear piezoelectricity are derived from Maxwell's equations, and Section 5.2.2 summarizes the initial-boundary value problem of linear piezoelectricity. Section 5.3 surveys the role of thermodynamics in the construction of various equivalent constitutive laws and their associated thermodynamic invariants. The structure of the constitutive laws generated by crystalline materials having different symmetry operations is described in Section 5.4.

Chapter 6 focuses on the use of Newton's method to derive the governing equations for linearly piezoelectric composite structures. The axial actuator model, which is a prototype for the linear actuators of the type depicted in Figure 1.8, is treated in Sections 6.1 and 6.2. Section 6.3 presents an analysis of the beam actuator as shown in the introduction in Figure 1.9. Section 6.4 uses Newton's method to derive the governing equations for a simplified model of piezoelectric composite plate bending.

Chapter 7 introduces powerful variational methods for deriving the governing equations of piezoelectric structures. The chapter begins with a review of variational calculus in Chapter 7.1. Hamilton's principle for mechanical systems is introduced in Section 7.2, and its generalization for linear piezoelectricity is presented in Section 7.3. The strength of these variational techniques is illustrated in Section 7.4, which shows how variational methods for electromechanical systems that consist of piezoelectric structures and attached ideal circuits can be modeled. Various authors have discussed variational methods for electromechanical systems over the years, and Section 7.5 discusses the relationships among some alternative forms of these principles. Section 7.6 illustrates how the electromechanical variational principle can be applied using Lagrangian densities $\mathfrak{L}$ instead of Lagrangian functions $\mathcal{L}$.

Chapter 8, the final chapter of this book gives a detailed description of approximation methods for linearly piezoelectric composite structures. The chapter begins in Section 8.1 with a discussion of the differences between classical, strong, and weak forms of the governing equations. As discussed in Section 8.1, the approximation strategies in the text are derived from the weak form of the governing equations. Section 8.2 gives a quick overview of modeling damping and dissipation, with the primary emphasis on viscous damping that is so popular in engineering vibrations texts. Galerkin approximations are introduced in Section 8.3, and example applications for the linear and bending actuators are summarized. Two classes of bases are described for use in the Galerkin approximations. Modal or eigenfunction bases are used in Section 8.3.1, and finite element functions are employed in Section 8.3.2. The chapter finishes with a collection of examples that summarize how transient and steady state solutions are obtained from the Galerkin approximations. Particular emphasis is placed on the derivation of complex frequency response equations and FRFs for the piezoelectric composite structures.

The Appendix contains three sections that provide supplementary material for the discussions throughout the text. Section S.1 gives a streamlined summary of the basic background for vibrations theory for single degree of freedom (SDOF) systems, distributed parameter systems (DPS), and multi-DOF (MDOF) systems. A supplementary account of tensor analysis is given in Section S.2. For those students seeking to understand the simplified version of tensor analysis covered in Chapter 2, this section shows how the simplified account fits in the general theory. Finally, Section S.3 discusses details regarding distributional and weak derivatives beyond the brief account in Chapter 8. A rigorous definition of a weak derivative is given, and the Sobolev spaces $W^{k,p}(\Omega)$ of all functions whose weak derivatives of order less than or equal to p are elements of the Lebesgue space $L^p(\Omega)$ is defined.

2

Mathematical Background

Piezoelectric materials exhibit coupling between their mechanical and electrical properties that is due to asymmetry in the underlying crystalline structure. These materials are anisotropic when considered at a macroscopic, continuum scale: their electrical and mechanical properties depend on direction. For these reasons it is crucial that we develop a mathematical description that is rich enough to characterize how the mechanical and electrical field variables transform with a change in coordinates. Section 2.1 describes how representations of vectors change when we vary the choice of basis. The transformation equations in this section are applicable to physical observables that are represented by vectors such as electric field, electric displacement, polarization, position, velocity, and acceleration. Section 2.2 generalizes this analysis and derives the transformation laws for n^{th} order tensors . The transformation laws in Section 2.2 are applicable to quantities such as the stress tensor, linear strain tensor, and the higher order tensors that appear in the linear piezoelectric constitutive laws in Chapter 5. Section 2.3 discusses how symmetry properties are described in terms of invariance under transformations, and how symmetry considerations manifest in tensor invariance.

2.1 Vectors, Bases, and Frames

A *dextral*, or *right-handed*, *coordinate frame* in $\mathbb{R}^3$ is defined in terms of three mutually orthogonal unit vectors $\mathbf{g}_i \in \mathbb{R}^3, i = 1, 2, 3,$ that permute cyclically according to the right hand rule under the cross product. The *orthonormality* condition is summarized as the requirement that

$$\mathbf{g}_i \cdot \mathbf{g}_j = \delta_{ij} := \begin{cases} 1 & \text{if } i = j, \\ 0 & \text{if } i \neq j, \end{cases}$$

Vibrations of Linear Piezostructures, First Edition. Andrew J. Kurdila and Pablo A. Tarazaga.

This Work is a co-publication between John Wiley & Sons Ltd and ASME Press.

where δ_{ij} is the *Kronecker delta function*. The requirement that the vectors permute cyclically according to the right hand rule is expressed succinctly in the following conditions:

$$\mathbf{g}_i = \begin{cases} \mathbf{g}_j \times \mathbf{g}_k & \text{if } (i,j,k) = (1,2,3), (2,3,1), \text{or } (3,1,2). \\ -\mathbf{g}_j \times \mathbf{g}_k & \text{if } (i,j,k) = (2,1,3), (1,3,2), \text{or } (3,2,1). \end{cases}$$

Figure 2.1 gives a graphical description of the *cyclic permutation condition* that is simple to remember. Such a set of vectors $\{\mathbf{g}_i\}_{i=1,2,3}$ that are orthonormal and permute cyclically under the cross product are said to form a *basis* for the coordinate frame. As shown in Figure 2.1, let $\boldsymbol{a} \in \mathbb{R}^3$ be an arbitrary vector in Cartesian space, and further suppose that $\{\mathbf{g}_i\}_{i=1,2,3}$ and $\{\mathbf{g}'_i\}_{i=1,2,3}$ are two different sets of right-handed orthonormal basis vectors. We can express the arbitrary vector $\boldsymbol{a}$ in terms of either basis as

$$\boldsymbol{a} = \sum_{i=1}^{3} a_i \mathbf{g}_i = \sum_{i=1}^{3} a'_i \mathbf{g}'_i \tag{2.1}$$

where a_i and a'_i are the components of the vector $\boldsymbol{a}$ relative to the bases $\{\mathbf{g}_i\}_{i=1,2,3}$ and $\{\mathbf{g}'_i\}_{i=1,2,3}$, respectively. Throughout this text we adopt the *Einsteinian summation convention* wherein the summation symbols in Equation 2.1 are suppressed and we write

$$\boldsymbol{a} = a_i \mathbf{g}_i = a'_i \mathbf{g}'_i. \tag{2.2}$$

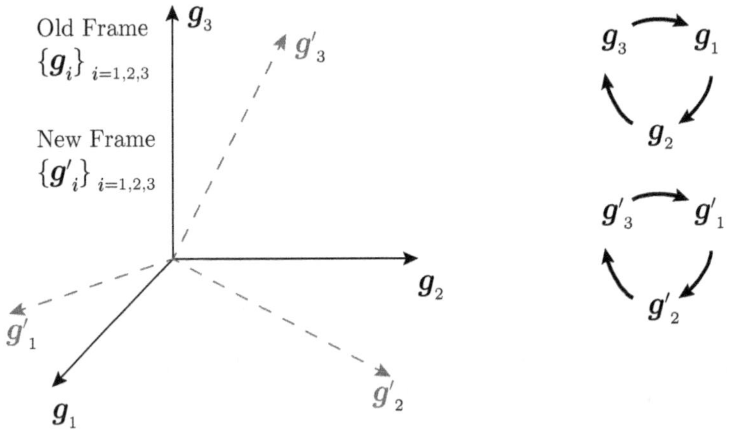

Figure 2.1 Frames generated by basis vectors $\mathbf{g}_i$ and $\mathbf{g}'_i$, $i = 1, 2, 3$, and their cyclic permutation diagrams for dextral bases.

In the summation convention, any repeated index in a product term corresponds to a summation. If we view $\{\mathbf{g}_i\}_{i=1,2,3}$ as defining an original set of coordinates, and interpret the frame generated by $\{\mathbf{g}'_i\}_{i=1,2,3}$ as corresponding to a new or transformed set of coordinates, we can solve for the new coordinates in terms of the old. We take the dot product of Equation 2.1 with an arbitrary basis vector $\mathbf{g}'_k$ and obtain $a_i\mathbf{g}'_k \cdot \mathbf{g}_i = a'_i\mathbf{g}'_k \cdot \mathbf{g}'_i = a'_i\delta_{ki} = a'_k$. We define the transformation $r_{ki} := \mathbf{g}'_k \cdot \mathbf{g}_i$ and the final form of the *change of coordinates* is given by

$$a'_k = r_{ki}a_i. \tag{2.3}$$

Recall that $\boldsymbol{a} \cdot \boldsymbol{b} = \|\boldsymbol{a}\|\|\boldsymbol{b}\| \cos(\theta_{\boldsymbol{a},\boldsymbol{b}})$ for any two vectors $\boldsymbol{a}$ and $\boldsymbol{b}$ where $\theta_{\boldsymbol{a},\boldsymbol{b}}$ is the angle between them. The entries r_{ki} are simply the *direction cosines* between the new $\mathbf{g}'_k$ axis and old $\mathbf{g}_i$ axis, respectively. In other words, we have $r_{ki} = \cos(\theta_{\mathbf{g}'_k,\mathbf{g}_i})$ where $\theta_{\mathbf{g}'_k,\mathbf{g}_i}$ is the angle between the axes $\mathbf{g}'_k$ and $\mathbf{g}_i$.

If we organize the entries r_{ki} in a matrix $\boldsymbol{R} := [r_{ki}]$ where k is the row and i is the column, the resulting matrix $\boldsymbol{R}$ is an *orthogonal matrix*, that is, $\boldsymbol{R}^T\boldsymbol{R} = \boldsymbol{R}\boldsymbol{R}^T = \boldsymbol{I}$. The matrix $\boldsymbol{R}$ is also known as a *direction cosine matrix* or a *rotation matrix*. The proof that $\boldsymbol{R}$ is an orthogonal matrix is constructive and provides a good example of the efficacy of calculations carried out using the *summation convention*.

Example 2.1.1 We begin by expanding Equation 2.1 for the specific choice $\boldsymbol{a} = \mathbf{g}'_k$

$$\mathbf{g}'_k = (\mathbf{g}'_k \cdot \mathbf{g}_j)\mathbf{g}_j,$$

and subsequently take the dot product of two such expressions for $\mathbf{g}'_k$ and $\mathbf{g}'_i$. When we use the orthonormality of the basis vectors and the properties of the Kronecker delta function, we see that

$$\begin{aligned} \delta_{ki} = \mathbf{g}'_k \cdot \mathbf{g}'_i &= ((\mathbf{g}'_k \cdot \mathbf{g}_j)\mathbf{g}_j) \cdot ((\mathbf{g}'_i \cdot \mathbf{g}_n)\mathbf{g}_n), \\ &= (\mathbf{g}'_k \cdot \mathbf{g}_j)(\mathbf{g}'_i \cdot \mathbf{g}_n)\delta_{jn}, \\ &= (\mathbf{g}'_k \cdot \mathbf{g}_j)(\mathbf{g}'_i \cdot \mathbf{g}_j) = r_{kj}r_{ij}. \end{aligned}$$

This sequence of calculations shows that $\delta_{ki} = r_{kj}r_{ij}$, and a similar line of reasoning shows that $\delta_{ki} = r_{jk}r_{ji}$. We conclude that the matrix form of these indicial expressions is

$$\boldsymbol{I} = \boldsymbol{R}^T\boldsymbol{R} = \boldsymbol{R}\boldsymbol{R}^T,$$

where $\boldsymbol{I}$ is the identity matrix, and $\boldsymbol{R}$ is the matrix whose entry in the jth row and k^{th} column is r_{jk}, $\boldsymbol{R} := [r_{jk}]$. In other words, $\boldsymbol{R}$ is an orthogonal matrix.

Further examples of the utility of the summation convention in computations related to continuum mechanics and electrodynamics are given in Problems 2.4.1 through 2.4.4.

2.2 Tensors

Equation 2.3 provides a complete description of how the components of some arbitrary physical vector $\boldsymbol{a}$ transform with a change of the frame of reference. It is valid for the vectors that represent position, displacement, velocity and acceleration as they are defined in continuum mechanics. It also holds for vectors that represent the electric field, electric displacement, and polarization that arise in the study of continuum electrodynamics or piezoelectricity. However, we require a generalization of this identity to understand how quantities such as stress, strain, or the constitutive laws of linear piezoelectricity transform with a change of coordinates. We must consider physically observable quantities that correspond to *tensors.*

Tensor analysis is presented and described fully in many texts on mathematics, mechanics, and differential geometry. A comprehensive account of the theory can be nuanced and subtle. Fortunately, only a very specialized treatment is required in this text. A thorough discussion can be found in [9, 33], or [52]. A brief introduction to the general theory is given in Appendix S.2.

The *tensor product* $\otimes$ is a binary operation on vector space V that defines elements of the *tensor product space* $V \otimes V$. Elements of $V \otimes V$ are examples of *second order tensors*. The tensor product operator $\otimes$ satisfies the following properties:

1. $\boldsymbol{u} \otimes (\boldsymbol{v}_1 + \boldsymbol{v}_2) = \boldsymbol{u} \otimes \boldsymbol{v}_1 + \boldsymbol{u} \otimes \boldsymbol{v}_2$ for all $\boldsymbol{u}, \boldsymbol{v}_1, \boldsymbol{v}_2 \in V$
2. $(\boldsymbol{u}_1 + \boldsymbol{u}_2) \otimes \boldsymbol{v} = \boldsymbol{u}_1 \otimes \boldsymbol{v} + \boldsymbol{u}_2 \otimes \boldsymbol{v}$ for all $\boldsymbol{u}_1, \boldsymbol{u}_2, \boldsymbol{v} \in V$
3. $\{\mathbf{g}_i\}_{i=1,2,3}$ is a basis for $V \implies \{\mathbf{g}_i \otimes \mathbf{g}_j\}_{i,j=1,2,3}$ is a basis for $V \otimes V$
4. $(\boldsymbol{u} \otimes \boldsymbol{v}) \cdot (\boldsymbol{x} \otimes \boldsymbol{y}) := (\boldsymbol{u} \cdot \boldsymbol{x})(\boldsymbol{v} \cdot \boldsymbol{y})$
5. $\boldsymbol{u} \cdot (\boldsymbol{x} \otimes \boldsymbol{y}) = (\boldsymbol{u} \cdot \boldsymbol{x})\boldsymbol{y}$ and $(\boldsymbol{x} \otimes \boldsymbol{y}) \cdot \boldsymbol{v} = \boldsymbol{x}(\boldsymbol{y} \cdot \boldsymbol{v})$

This abstract definition can be made concrete by considering the following example.

Example 2.2.1 Let $V = \mathbb{R}^3$, $\boldsymbol{u} := 3\mathrm{g}_1 + 4\mathrm{g}_2 + 5\mathrm{g}_3$, and $\boldsymbol{v} := 1\mathrm{g}_1 + 2\mathrm{g}_2 + 3\mathrm{g}_3$. In this case we have

$$\begin{aligned} \boldsymbol{u} \otimes \boldsymbol{v} &= (3\mathrm{g}_1 + 4\mathrm{g}_2 + 5\mathrm{g}_3) \otimes (1\mathrm{g}_1 + 2\mathrm{g}_2 + 3\mathrm{g}_3), \\ &= 3\mathrm{g}_1 \otimes \mathrm{g}_1 + 6\mathrm{g}_1 \otimes \mathrm{g}_2 + 9\mathrm{g}_1 \otimes \mathrm{g}_3 \end{aligned}$$

$$+ 4\mathbf{g}_2 \otimes \mathbf{g}_1 + 8\mathbf{g}_2 \otimes \mathbf{g}_2 + 12\mathbf{g}_2 \otimes \mathbf{g}_3$$
$$+ 5\mathbf{g}_3 \otimes \mathbf{g}_1 + 10\mathbf{g}_3 \otimes \mathbf{g}_2 + 15\mathbf{g}_3 \otimes \mathbf{g}_3.$$

Note that, when we use the summation convention, we can write

$$\boldsymbol{u} \otimes \boldsymbol{v} = t_{ij}\mathbf{g}_i \otimes \mathbf{g}_j,$$

where the components t_{ij} can be organized as a matrix

$$[t_{ij}] = \begin{bmatrix} 3 & 6 & 9 \\ 4 & 8 & 12 \\ 5 & 10 & 15 \end{bmatrix}.$$

The collection of constants t_{ij} are the components of the tensor $\boldsymbol{t} = \boldsymbol{u} \otimes \boldsymbol{v}$ with respect to the tensor product basis $\{\mathbf{g}_i \otimes \mathbf{g}_j\}_{i,j=1,2,3}$ for $V \otimes V$.

From the properties of the tensor product operator $\otimes$, we can show that if $\{\mathbf{g}_i\}_{i=1,2,3}$ and $\{\mathbf{g}'_i\}_{i=1,2,3}$ are bases for V, then $\{\mathbf{g}_i \otimes \mathbf{g}_j\}_{i,j=1,2,3}$ and $\{\mathbf{g}'_i \otimes \mathbf{g}'_j\}_{i,j=1,2,3}$ are bases for $V \otimes V$ that satisfy

$$(\mathbf{g}'_i \otimes \mathbf{g}'_j) \cdot (\mathbf{g}'_m \otimes \mathbf{g}'_n) = \delta_{im}\delta_{jn} = (\mathbf{g}_i \otimes \mathbf{g}_j) \cdot (\mathbf{g}_m \otimes \mathbf{g}_n).$$

Any arbitrary tensor $\boldsymbol{t} \in V \otimes V$ can be written in terms of components t_{ij} and t'_{ij} relative to either basis

$$\boldsymbol{t} = t_{ij}\mathbf{g}_i \otimes \mathbf{g}_j = t'_{ij}\mathbf{g}'_i \otimes \mathbf{g}'_j. \tag{2.4}$$

Following the reasoning that is analogous to that used in Section 2.1, it is possible to solve for the transformed or new components t'_{ij} of the tensor $\boldsymbol{t}$ in terms of the original components t_{ij} by taking the dot product of Equation 2.4 with an arbitrary basis element $\mathbf{g}'_m \otimes \mathbf{g}'_l$.

$$\underbrace{t'_{ij}\underbrace{(\mathbf{g}'_m \otimes \mathbf{g}'_l) \cdot (\mathbf{g}'_i \otimes \mathbf{g}'_j)}_{\delta_{im}\delta_{jl}}}_{t'_{ml}} = t_{ij}\underbrace{(\mathbf{g}'_m \otimes \mathbf{g}'_l) \cdot (\mathbf{g}_i \otimes \mathbf{g}_j)}_{\underbrace{(\mathbf{g}'_m \cdot \mathbf{g}_i)(\mathbf{g}'_l \cdot \mathbf{g}_j)}_{r_{mi}r_{lj}}}$$

The resulting *transformation law* for *second order tensors* is

$$t'_{ml} = r_{mi}r_{lj}t_{ij},$$

which is structurally similar to the transformation law in Equation 2.3 for vectors.

Finally, the procedure used to derive transformation laws for second order tensors in $V \otimes V$ can be generalized to define transformation laws for order n tensors in $V \otimes V \otimes \cdots \otimes V$. Each order n tensor $\boldsymbol{t} \in V \otimes V \otimes \cdots \otimes V$ can be written in either form as

$$\begin{aligned} \boldsymbol{t} &= t_{i_1 i_2 \cdots i_n} \mathbf{g}_{i_1} \otimes \mathbf{g}_{i_2} \otimes \cdots \otimes \mathbf{g}_{i_n}, \\ &= t'_{i_1 i_2 \cdots i_n} \mathbf{g}'_{i_1} \otimes \mathbf{g}'_{i_2} \otimes \cdots \otimes \mathbf{g}'_{i_n}, \end{aligned}$$

where the components can be shown to transform under a change of basis according to the transformation law

$$t'_{i_1 i_2 \cdots i_n} = r_{i_1 j_1} r_{i_2 j_2} \cdots r_{i_n j_n} t_{j_1 j_2 \cdots j_n}.$$

Some authors use this transformation rule to define an nth order tensor. See [15] for an example in the field of continuum mechanics.

The following example is a typical application of tensor calculations in mechanics and illustrates the utility of the transformation laws in applications.

Example 2.2.2 The *rotational kinetic energy* T_ω of a rigid body is often written in the form

$$T_\omega = \frac{1}{2} \boldsymbol{\omega} \cdot \Box \cdot \boldsymbol{\omega},$$

where $\boldsymbol{\omega}$ is the *angular velocity* vector of the rigid body and $\Box$ is the second order inertia tensor about the body center of mass. In this example we find the matrix form of the transformation law that expresses how the components of the inertia tensor transform with a rotation of coordinate system. We also derive equivalent expressions that can be used to compute the rotational kinetic energy from the components of the angular velocity vector and the inertia tensor.

Suppose that $\{\mathbf{g}_i\}_{i=1,2,3}$ and $\{\mathbf{g}'_i\}_{i=1,2,3}$ are the bases of two different coordinate systems. Since $\boldsymbol{\omega}$ and $\Box$ are examples of first and second order tensors, respectively, we can expand them as

$$\begin{aligned} \boldsymbol{\omega} &= \omega_i \mathbf{g}_i = \omega'_i \mathbf{g}'_i, \\ \Box &= \Box_{kl} \mathbf{g}_k \otimes \mathbf{g}_l = \Box'_{kl} \mathbf{g}'_k \otimes \mathbf{g}'_l. \end{aligned} \tag{2.5}$$

We can solve for the components $\Box'_{kl}$ of the inertia relative to the basis $\{\mathbf{g}'_k \otimes \mathbf{g}'_l\}_{i,j=1,2,3}$ in terms of the components $\Box_{kl}$ relative to the basis $\{\mathbf{g}_k \otimes \mathbf{g}_l\}_{i,j=1,2,3}$ by taking the dot product of Equation 2.5 with an arbitrary basis element $\mathbf{g}'_m \otimes \mathbf{g}'_n$.

$$\begin{aligned} \Box'_{kl} (\mathbf{g}'_m \otimes \mathbf{g}'_n) \cdot (\mathbf{g}'_k \otimes \mathbf{g}'_l) &= \Box_{kl} (\mathbf{g}'_m \otimes \mathbf{g}'_n) \cdot (\mathbf{g}_k \otimes \mathbf{g}_l), \\ \Box'_{kl} \delta_{mk} \delta_{nl} &= r_{mk} r_{nl} \Box_{kl}, \\ \Box'_{mn} &= r_{mk} r_{nl} \Box_{kl}. \end{aligned}$$

We can now compute the rotational kinetic energy T_ω in terms of components relative to either set of bases. For example, we have

$$\begin{aligned} T_\omega &= \frac{1}{2}(\omega_i \mathbf{g}_i) \cdot (\mathbb{I}_{kl}\mathbf{g}_k \otimes \mathbf{g}_l) \cdot (\omega_j \mathbf{g}_j), \\ &= \frac{1}{2}\omega_i \mathbb{I}_{kl} \omega_j (\mathbf{g}_i \cdot \mathbf{g}_k)(\mathbf{g}_j \cdot \mathbf{g}_l), \\ &= \frac{1}{2}\omega_i \mathbb{I}_{kl} \omega_j \delta_{ik} \delta_{jl} = \frac{1}{2}\omega_k \mathbb{I}_{kl} \omega_l. \end{aligned}$$

Alternatively, we can calculate the rotational kinetic energy in terms of components relative to the transformed basis to find that

$$\begin{aligned} T_\omega &= \frac{1}{2}(\omega'_i \mathbf{g}'_i) \cdot (\mathbb{I}'_{kl}\mathbf{g}'_k \otimes \mathbf{g}'_l) \cdot (\omega'_j \mathbf{g}'_j), \\ &= \frac{1}{2}\omega'_k \mathbb{I}'_{kl} \omega'_l. \end{aligned}$$

We can likewise show that the two expressions are equivalent since we have

$$\begin{aligned} T_\omega = \frac{1}{2}\omega'_k \mathbb{I}'_{kl} \omega'_l &= \frac{1}{2} r_{ks}\omega_s \mathbb{I}'_{kl} r_{lt}\omega_t, \\ &= \frac{1}{2}\omega_s (r_{ks}\mathbb{I}'_{kl} r_{lt})\omega_t = \frac{1}{2}\omega_s \mathbb{I}_{st}\omega_t. \end{aligned}$$

The last line above follows from the *second order tensor transformation law* for $\mathbb{I}'_{kl}$.

Example 2.2 illustrates the utility of the summation convention in calculations involving tensors and their transformation rules, and the ease with which the transformation rules can be cast as matrix equations for first or second order tensors. This text will make widespread use of both the full vector notation and the tensor notation that exploits the summation convention, depending on which is more convenient in a specific equation, problem, or application. The problems at the end of this chapter should afford the student the opportunity to become proficient in both representations, and to be able to pass easily from one format to another. In many cases we do not explicitly label the bases used for representation of a tensor, nor do we explicitly form an equivalent matrix representation. We often just list the components relative to some implied basis. That is, we simply refer to the tensor $t_{i_1,i_2,\ldots,i_n}$ and refrain from the explicit formula

$$\boldsymbol{t} = t_{i_1,i_2,\ldots,i_n} \mathbf{g}_{i_1} \otimes \mathbf{g}_{i_2} \otimes \cdots \otimes \mathbf{g}_{i_n}.$$

For example, in our study of the fundamentals of continuum mechanics in Chapter 3, we refer to the second order stress tensor $\boldsymbol{T} = T_{ij}\mathbf{g}_i \otimes \mathbf{g}_j$ simply as T_{ij}. Similarly, we refer to the *second order linear strain tensor* $\boldsymbol{S} = S_{ij}\mathbf{g}_i \otimes \mathbf{g}_j$ as S_{ij}. In our discussion of continuum electrodynamics in Chapter 4 we refer to the *electric field*

$\boldsymbol{E} := E_i \mathbf{g}_i$ as E_i and the *electric displacement* $\boldsymbol{D} := D_i \mathbf{g}_i$ as D_i, respectively. Our discussion of the constitutive laws for linearly piezoelectric materials in Chapter 5 refers to the third order *piezoelectric coupling tensor* $\mathbf{e} := e_{ijk} \mathbf{g}_i \otimes \mathbf{g}_j \otimes \mathbf{g}_k$ as e_{ijk} and to the *fourth order stiffness tensor* $C^E := C^E_{ijkl}\, \mathbf{g}_i \otimes \mathbf{g}_j \otimes \mathbf{g}_k \otimes \mathbf{g}_l$ as C^E_{ijkl}.

2.3 Symmetry, Crystals, and Tensor Invariance

The vector and *tensor transformation rules* discussed in Sections 2.1 and 2.2 play a crucial role in understanding material symmetry. They are thereby fundamental in particular to understanding the physics of piezoelectric materials.

2.3.1 Geometry of Crystals

Materials with a *crystalline structure* are those in which the constituent atoms, molecules, or ions are organized in a regular array that is composed of identical, repeating *unit cells*. Discussions of the macroscopic piezoelectric material properties often explain symmetry in terms of the location of molecules or ions; we will simply refer to ion locations in the discussions that follow. A typical unit cell is depicted in Figure 2.2, and a closeup of the specific unit cell for the class of *triclinic materials* is shown in Figure 2.3. The geometry of the unit cell is characterized by the *lattice vectors* $\boldsymbol{a}, \boldsymbol{b}, \boldsymbol{c}$ that are aligned with the edges of the unit cell and with the *lattice parameters*. The lattice parameters include the lengths a, b, c of the lattice vectors $\boldsymbol{a}, \boldsymbol{b}, \boldsymbol{c}$ and the angles α, β, γ that measure between the $(\boldsymbol{b}, \boldsymbol{c}), (\boldsymbol{a}, \boldsymbol{c}), (\boldsymbol{a}, \boldsymbol{b})$ axes, respectively.

All unit cells contain ions that are located at the corners of the unit cell. Such a cell is called a *primitive unit cell*. A unit cell that in addition contains ions on two opposing faces of the unit cell is said to be *side centered*. A *body centered* unit cell contains ions at the cell corners and within the body of the cell, while a *face centered* cell contains ions at the corners and all of the faces of the cell.

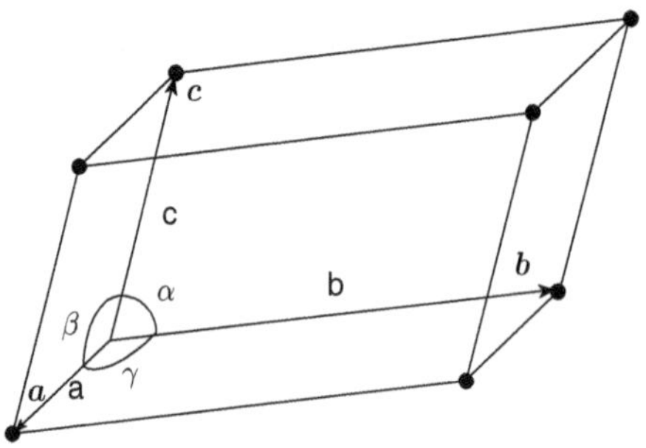

Figure 2.2 Unit cell and lattice parameters.

Figure 2.3 The unit cell of the triclinic unit cell.

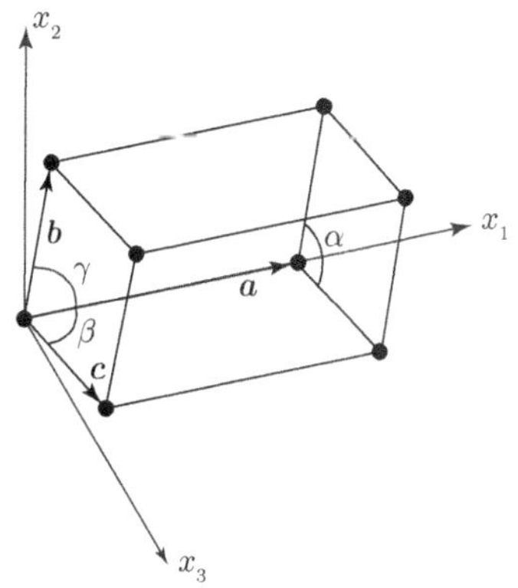

The crystalline *lattice* is defined by the collection of points in three dimensions that are occupied by ions when the unit cells are assembled to form a material continuum. Coordinates within the lattice are specified in terms of integer multiples of the lattice parameters. That is, a triple of integers $(\alpha_1, \alpha_2, \alpha_3)$ represents the vector that locates the lattice site at $\alpha_1\boldsymbol{a} + \alpha_2\boldsymbol{b} + \alpha_3\boldsymbol{c}$. It is also conventional in crystallographic notation to use the overbar to designate a negative, as in $(\overline{\alpha_1}, \alpha_2, \alpha_3) = (-\alpha_1, \alpha_2, \alpha_3)$. Crystallographic directions are designated using the square brackets containing integer expressions such as $[\beta_1, \beta_2, \beta_3]$. By convention the integers $\beta_1, \beta_2, \beta_3$ must not contain a common integer factor, other than one. A plane that passes through three non-collinear lattice sites defines a Crystallographic plane, which is defined in terms of Miller indices *mno*. The *Miller indices mno* of a crystalline plane are defined to be the reciprocal of the intercepts of the plane with the crystallographic lattice vectors $\boldsymbol{a}, \boldsymbol{b}, \boldsymbol{c}$. In other words, the coordinates of the intercepts of the crystallographic plane with the lattice vector directions are $(1/m, 1/n, 1/o)$. As in the case of crystallographic directions, the integers in the Miller indices must be relatively prime: they do not possess a common factor other than one. A schematic representation of crystallographic coordinates, directions, and planes is given in Figure 2.4.

Because of the requirement that the unit cells when assembled must conform to a regular array, one that is translation invariant, there are only fourteen possible unit cells and associated lattice structures in three dimensions. These fourteen possible arrangements define the *Bravais lattices* which are depicted in Figure 2.5. This collection is further subdivided into the seven *crystal systems*: *cubic*, *tetragonal*, *orthorhombic*, *monoclinic*, *hexagonal*, *trigonal*, and *triclinic*. Each crystal system includes a primitive unit cell, and may also include side centered, face centered, or body centered unit cells. The defining constraints on the unit cell for the seven crystal systems are tabulated in Figure 2.5. The triclinic unit cell has the form of a parallelopiped with $a \neq b \neq c$ and $\alpha \neq \beta \neq \gamma \neq \pi/2$. A monoclinic material has a

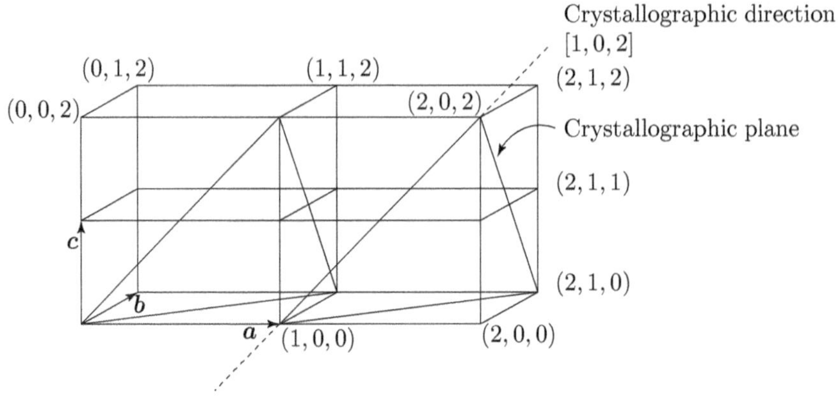

Figure 2.4 Crystallographic coordinates, directions, and planes.

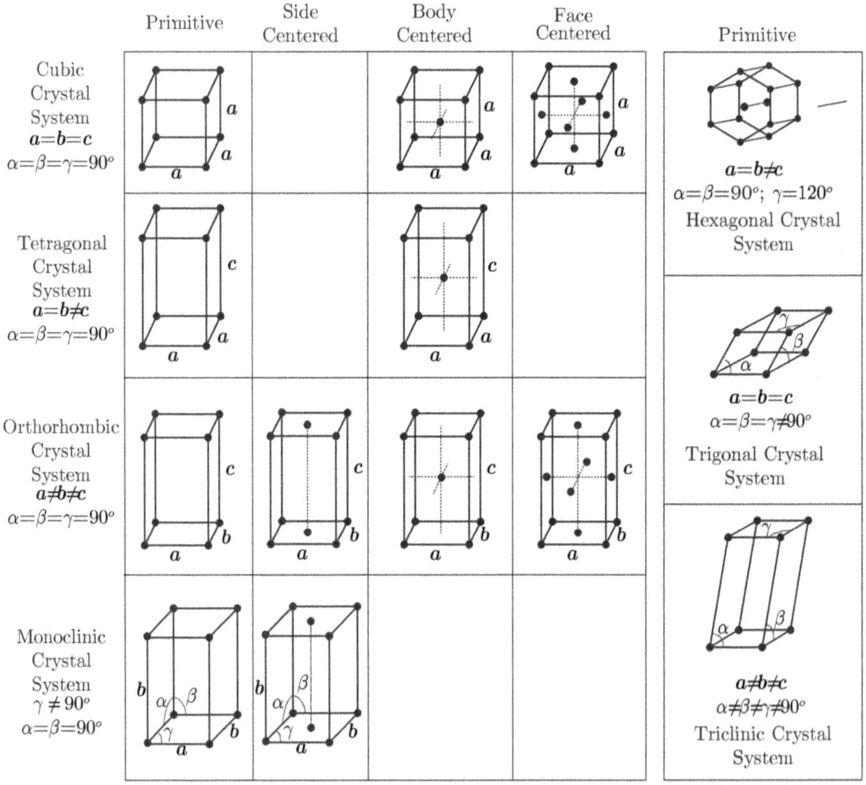

Figure 2.5 The fourteen Bravais lattices, seven crystal systems, and associated unit cells. See [43] or [23]. Source: Modified from Jan Tichy, Jiri Erhart, Erwin Kittinger, and Jana Privratska, Fundamentals of Piezoelectric Sensorics, SpringerVerlag, Inc., Berlin, 2010; Maureen M. Julian, Foundations of Crystallography, CRC Press, 2008.

unit cell characterized by $\alpha = \beta = \pi/2$ and $a \neq b \neq c$. The base of the unit cell of a monoclinic material is a therefore a parallelogram in the $\boldsymbol{a}, \boldsymbol{b}$ plane with included angle $\gamma \neq \pi/2$. The base parallelogram is extruded along the perpendicular $\boldsymbol{c}$ vector to generate the solid. An orthorhombic unit cell is defined by the constraints that $\alpha = \beta = \gamma = \pi/2$ and $a \neq b \neq c$. A tetragonal unit cell must have $\alpha = \beta = \gamma = \pi/2$, but $a = b \neq c$. The cubic unit cell has $a = b = c$ and $\alpha = \beta = \gamma = \pi/2$ by definition. The trigonal or rhombohedral unit cell has equal length sides and angles, so that $a = b = c$ and $\alpha = \beta = \gamma$. Finally, the hexagonal unit cell has base angle $\gamma = 120°$ and side angles $\alpha = \beta = \pi/2$, while the side lengths satisfy $a = b \neq c$.

The symmetry properties at a macroscopic level of crystalline materials are described in a precise way by tabulating the *symmetry operations* that leave the unit cell or crystal lattice invariant. A symmetry operation for an object is a transformation that maps the object into one that has the same appearance. Symmetry operations can be grouped into those that describe the symmetry of the lattice under translation and those that express the symmetry at a certain point in the lattice. These latter symmetry operations are known as *point symmetry operations*. Common symmetry transformations include rotation about an axis of symmetry, reflection about a plane of symmetry, inversion, or a combination of these operations. A reflection is sometimes referred to as a *mirror transformation*. The inversion symmetry operation, which is symbolically represented as $\bar{1}$, maps a particular point across an inversion point. If the inversion point is defined to be the origin, the inversion mapping assigns the point (a, b, c) to $(\bar{a}, \bar{b}, \bar{c})$. Symmetries in crystallography are also often described in terms of *rotoinversions*, a rotation followed by an inversion [28]. An example of a rotoinversion that consists of a rotation of $\pi/2$ followed by an inversion about the origin is depicted in Figure 2.6. An object exhibits an n-fold rotational symmetry is there is an axis about which a rotation of $360°/n$ maps the object into one having the same appearance. In three dimensions, it is possible to define two, three, four, and sixfold symmetry about an axis. Figure 2.7 illustrates examples of two, three, and fourfold axes of symmetry.

The collection of all possible symmetry transformations for a given crystal lattice can be large, and the determination of which symmetry transformations are associated with a particular crystal can lead to a subtle analysis. A comprehensive presentation is beyond the scope of this book. A good review and summary can be found in Appendix A of [13]. References [28] and [23] provide a detailed and systematic discussion of crystal symmetry from first principles. [13] includes an excellent discussion of tensor transformations and invariance of constitutive laws. In this textbook, we focus on general symmetry classes pertinent to linear elasticity in Chapter 3 and on a few common crystal structures in Chapter 5 that appear in the more familiar piezoelectric materials. For example, samples of Y-cut quartz are natural piezoelectric materials that belong to the monoclinic crystal system.

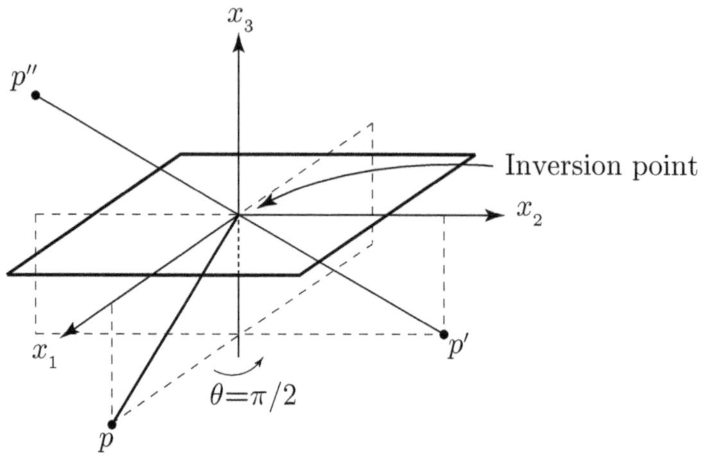

Figure 2.6 Rotoinversion p'' of the point p. The rotation of $\pi/2$ is followed by an inversion about the origin.

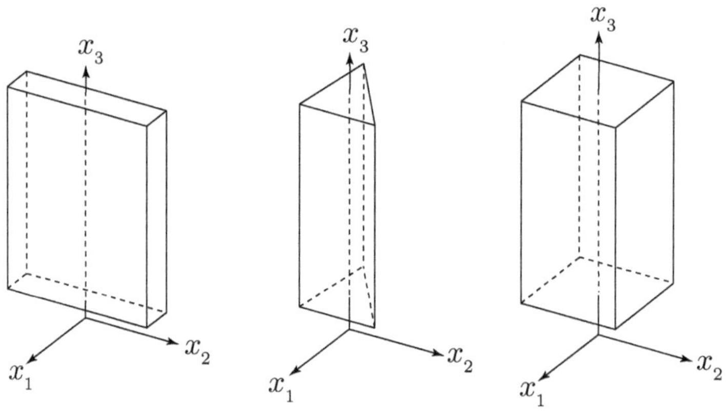

Figure 2.7 Examples of objects with two, three, and fourfold symmetry about the x_3 axis.

Polarized ferroelectric ceramics are another increasingly common piezoelectric material, ones that exhibit the symmetry of the hexagonal crystal system [44].

While the entire collection of symmetry operations associated with a given crystal will include translations and point symmetry operations, a finer description of symmetry into 32 *crystal classes* is obtained by further subdividing the seven crystal systems based on their point symmetry transformations. Each of the 32 crystal classes is associated with one of the 32 *crystallographic point groups* in three dimensions: the *point group* is the precise mathematical description of point symmetry

Table 2.1 The crystal systems and their point groups. See [13] Table A.2 or [43] Tables 2.1 and 2.2. Source: Modified from A. Cemal Eringen and Gerard A Maugin, Electrodynamics of Continua I, Springer Verlag, Inc., New York, 1990.; Jan Tichy, Jiri Erhart, Erwin Kittinger, and Jana Privratska, Fundamentals of Piezoelectric Sensorics, Springer Verlag, Inc., Berlin, 2010.

Crystal System	Point group	Point group
	Schoenflies	Hermann–Mauguin
Triclinic	C_1, C_i	$1, \bar{1}$
Monoclinic	C_2, C_s, C_{2h}	$2, m, 2/m$
Orthorhombic	D_2, D_{2v}, D_{2h}	$222, mm2, mmm$
Tetragonal	$C_4, S_4, C_{4h}, D_4,$	$4, \bar{4}, 4/m,$
	C_{4v}, D_{2d}, D_{4h}	$422, \bar{4}2m, 4/mmm$
Trigonal	$C_3, C_{3i}, D_3, C_{3v}, D_{3d}$	$3, \bar{3}, 32, 3m, \bar{3}m$
Hexagonal	$C_6, C_{3h}, C_{6h}, D_6,$	$6, \bar{6}, 6/m, 622,$
	C_{6v}, D_{3h}, D_{6h}	$6mm, \bar{6}m2, 6/mmm$
Cubic	T, T_h, O, T_d, O_h	$23, m\bar{3}, 432, \bar{4}3m, m\bar{3}m$

operations that characterize the crystal class. A three dimensional point group is a collection of operations that map at least one point of an object into itself [23]. A point group is commonly designated by its International Symbol that ranges from 1 to 32, its Schoenfliess Symbol, or its Hermann–Mauguin symbol. The point groups for the seven crystal classes are listed in Table 2.1, along with their Schoenfliess and Hermann–Mauguin symbols.

Table A.2 in [13] is particularly useful since it gives a concise list of the point symmetry transformations for all the unit cells in the seven crystal systems. In this text we will see, in Chapters 4–7, that the monoclinic and hexagonal systems are more relevant to our applications of linearly piezoelectric systems. The following example illustrates some of the geometric properties of the monoclinic crystal system, which are representative of the computations required in later chapters.

Example 2.3.1 With our focus on linearly piezoelectric materials, we can learn much about material symmetry by studying a few specific point symmetry operations. Consider, for example, a material which belongs to the monoclinic crystal system. The information in the table below can be found

(Continued)

Example 2.3.1 (Continued)

in Table A.2 in [13], or in the text [53] and summarizes the crystal classes that constitute this crystal system.

System	Class Number	Class Name	Designation	Symmetry
Monoclinic	3	Sphenoidal	C_2 or 2	twofold axis
	4	Domatic	C_s or m	symmetry plane
	5	Prismatic	C_{2h} or $2/m$	twofold axis, symmetry plane, inversion about origin

Let us verify the invariance of the lattice for a monoclinic material of class C_s or m. Suppose, specifically, that the $x_2 - x_3$ plane is a plane of material symmetry for such a crystal system, as shown in Figure 2.8.

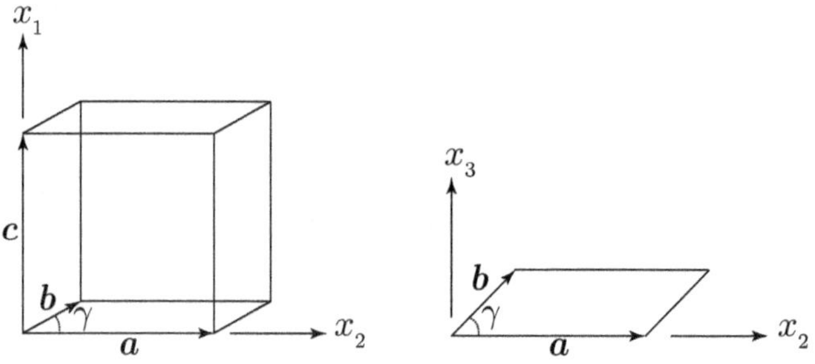

Figure 2.8 Unit cell of monoclinic crystal system.

In this case the crystal lattice should be invariant with respect to a reflection of the x_1 axis. The reflection about the plane whose normal is the x_1 axis is accomplished via the (improper) rotation matrix

$$r_{ij} = \begin{bmatrix} -1 & 0 & 0 \\ 0 & 1 & 0 \\ 0 & 0 & 1 \end{bmatrix}.$$

In this simple example, invariance with respect to reflection means that the points of the crystal lattice will be mapped into points of the lattice. If x_{pi} are the coordinates of a lattice point p, it must be the case that the coordinates x'_{pi} of the lattice point p with respect to the transformed basis

$$x'_{pi} = r_{ij} x_{pj},$$

again are a point of the lattice. It is evident from Figure 2.9 that the points p, q, r, s of the unit cell are mapped under the reflection r_{ij} to points p', q', r', s' on the monoclinic lattice.

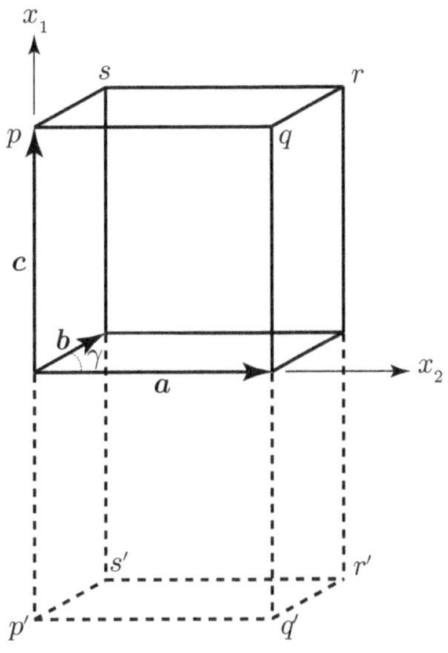

Figure 2.9 Invariance of the monoclinic lattice with respect to reflection about the $x_2 - x_3$ plane.

We next consider the crystal class C_2 or 2 and show that its lattice is invariant with respect to rotation through π about the two-fold axis of symmetry. The monoclinic unit cell in Figure 2.9 is symmetric with respect to a two-fold rotation about the x_1 axis. The rotation matrix associated with this symmetry transformation is

$$r_{ij} = \begin{bmatrix} 1 & 0 & 0 \\ 0 & -1 & 0 \\ 0 & 0 & -1 \end{bmatrix}.$$

(Continued)

Example 2.3.1 (Continued)

The points p, q, r on the unit cell in Figure 2.10 are mapped under this rotation matrix to the points p', q', r' on the monoclinic lattice, and the invariance of the lattice is clear.

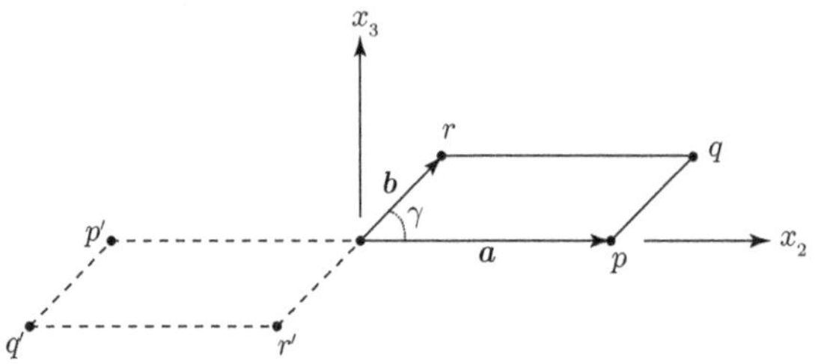

Figure 2.10 Invariance of the monoclinic lattice with respect to rotation through π about the two-fold x_1 axis.

We consider these two classes of crystal again when we discuss the invariance of constitutive laws in Sections 3.4 and 5.3.

2.3.2 Symmetry of Tensors

We have seen that when we perform a change of variables defined by the collection of transformations $r_{i_1j_1}, \ldots, r_{i_nj_n}$, we calculate the components $t'_{i_1i_2\ldots i_n}$ of an nth order tensor relative to the transformed coordinates in terms of the components $t_{j_1j_2\ldots j_n}$ relative to the original coordinates via the equation

$$t'_{i_1i_2\ldots i_n} = r_{i_1j_1}r_{i_2j_2}\cdots r_{i_nj_n}t_{j_1j_2\ldots j_n}.$$

In our study of the constitutive laws of piezoelectric materials, we will seek to understand the special structure that tensors exhibit when we enforce constraints on the symmetry of the material. By definition, an nth order tensor is said to be symmetric with respect to a set of transformations $r_{i_1j_1}, \ldots, r_{i_nj_n}$ whenever

$$t'_{i_1i_2\ldots i_n} = t_{i_1i_2\ldots i_n} = r_{i_1j_1}r_{i_2j_2}\cdots r_{i_nj_n}t_{j_1j_2\ldots j_n}.$$

This characterization of the invariance of an nth order tensor under a collection of symmetry transformations will prove to be essential in determining the form of constitutive laws for linearly piezoelectric materials in Chapter 5. For example, we show in Section 5.1 that the constitutive laws for linearly piezoelectric materials can be constructed from a fourth order stiffness tensor, a third order piezoelectric coupling tensor, and a second order permittivity tensor. The following example illustrates how the invariance of a second order tensor can be used to deduce its structure.

Example 2.3.2 Suppose that we have a second order tensor t_{ij} that is symmetric with respect to the symmetry transformations that characterize a monoclinic material with point symmetry of type C_2 or 2. We can infer much about the structure of the tensor by noting that we must have

$$t'_{ij} = t_{ij} = r_{im}r_{jn}t_{mn},$$

for the rotation matrix

$$r_{ij} := \begin{bmatrix} 1 & 0 & 0 \\ 0 & -1 & 0 \\ 0 & 0 & -1 \end{bmatrix}.$$

When we enforce the only nonzero terms from the rotation matrix in the summation, we find the following conditions:

$$\begin{aligned}
t_{11} &= r_{11}r_{11}t_{11} = (+1)(+1)t_{11} = t_{11},\\
t_{12} &= r_{11}r_{22}t_{12} = (+1)(-1)t_{12} = -t_{12},\\
t_{13} &= r_{11}r_{33}t_{13} = (+1)(-1)t_{13} = -t_{13},\\
t_{21} &= r_{22}r_{11}t_{21} = (-1)(+1)t_{21} = -t_{21},\\
t_{22} &= r_{22}r_{22}t_{22} = (-1)(-1)t_{22} = t_{22},\\
t_{23} &= r_{22}r_{33}t_{23} = (-1)(-1)t_{23} = t_{23},\\
t_{31} &= r_{33}r_{11}t_{31} = (-1)(+1)t_{31} = -t_{31},\\
t_{32} &= r_{33}r_{22}t_{32} = (-1)(-1)t_{32} = t_{32},\\
t_{33} &= r_{33}r_{33}t_{33} = (-1)(-1)t_{33} = t_{33}.
\end{aligned}$$

We conclude that when the second order tensor t_{ij} is invariant with respect to the particular transformation r_{ij} listed above, it must be true that

$$t_{12} = t_{13} = t_{21} = t_{31} = 0.$$

We can picture the constraints more readily by organizing the entries t_{ij} in the two dimensional array

$$t_{ij} := \begin{bmatrix} t_{11} & 0 & 0 \\ 0 & t_{22} & t_{23} \\ 0 & t_{32} & t_{33} \end{bmatrix}.$$

More examples of the relationship between symmetry, transformation laws, and tensors are covered in the problems at the end of this chapter, Chapter 3, and Chapter 5.

2.4 Problems

Problems 2.4.1 The Levi-Civita permutation symbol ϵ_{ijk} satisfies

$$\epsilon_{ijk} = \begin{cases} 1 & \text{if } (i,j,k) = (1,2,3), (2,3,1), \text{or } (3,1,2), \\ -1 & \text{if } (i,j,k) = (2,1,3), (1,3,2), \text{or } (3,2,1), \\ 0 & \text{otherwise.} \end{cases}$$

Show the following:

1. $\epsilon_{ijk}\epsilon_{jki} = 6$
2. $\epsilon_{ijk}A_jA_k = 0$
3. If $\boldsymbol{w} = w_i\mathbf{g}_i$, $\boldsymbol{u} = u_i\mathbf{g}_i$, $\boldsymbol{v} = v_i\mathbf{g}_i$, and $\boldsymbol{w} = \boldsymbol{u} \times \boldsymbol{v}$, then $w_i = \epsilon_{ijk}u_jv_k$.

Problems 2.4.2 Use indicial notation to prove the following identities

1. $\boldsymbol{u} \times (\boldsymbol{v} \times \boldsymbol{w}) = (\boldsymbol{u} \cdot \boldsymbol{w})\boldsymbol{v} - (\boldsymbol{u} \cdot \boldsymbol{v})\boldsymbol{w}$
2. $\boldsymbol{u} \times \boldsymbol{v} = -\boldsymbol{v} \times \boldsymbol{u}$
3. $(\boldsymbol{s} \times \boldsymbol{t}) \cdot (\boldsymbol{u} \times \boldsymbol{v}) = (\boldsymbol{s} \cdot \boldsymbol{u})(\boldsymbol{t} \cdot \boldsymbol{v}) - (\boldsymbol{s} \cdot \boldsymbol{v})(\boldsymbol{t} \cdot \boldsymbol{u})$
4. $\nabla \times (\nabla \times \boldsymbol{v}) = \nabla(\nabla \cdot \boldsymbol{v}) - \Delta \boldsymbol{v}$

Problems 2.4.3 Suppose that the components of the second order tensors $\boldsymbol{u}$ and $\boldsymbol{v}$ satisfy a linear relationship that has the form

$$u_{ij} = c_{ijkl}v_{kl}$$

when the components u_{ij} and v_{ij} are defined with respect to the basis $\{\mathbf{g}_i\}_{i=1,2,3}$. Use indicial notation to show that the coefficients c_{ijkl} obey the transformation laws of a fourth order tensor. In other words, if $\{\mathbf{g}'_i\}_{i=1,2,3}$ is a different basis, and the linear relationship has the form

$$u'_{ij} = c'_{ijkl}v'_{kl}$$

with respect to this new basis, then it must be the case that

$$c'_{ijkl} = r_{im}r_{jn}r_{ko}r_{lp}c_{mnop}.$$

Problems 2.4.4 Express the following well-known identities from vector calculus in indicial notation:

$$\nabla \times \nabla a = 0 \quad \text{for all smooth scalar functions } a, \text{ and}$$

$$\nabla \cdot \nabla \times \boldsymbol{A} = 0 \quad \text{for all smooth vector functions } \boldsymbol{A}.$$

Problems 2.4.5 Express Gauss's Theorem below in terms of indicial notation:

$$\int_{\Omega} \nabla \cdot \boldsymbol{A} dv = \int_{\partial\Omega} \boldsymbol{A} \cdot \boldsymbol{n} da$$

Problems 2.4.6 Determine the structure of the third order tensor t_{ijk} if it is invariant with respect to the symmetry transformations that characterize a monoclinic material of the C_2 or 2 crystal class.

Problems 2.4.7 Determine the structure of a third order tensor t_{ijk} if it is invariant with respect to the symmetry transformations that characterize a monoclinic material of the C_s or m crystal class.

Problems 2.4.8 Find a matrix that represents a two-fold symmetry around the x_2 axis.

Problems 2.4.9 Find a matrix that represents a three-fold symmetry about the x_3 axis.

Problems 2.4.10 Find a matrix that represents a four fold-symmetry about the x_3 axis.

Problems 2.4.11 Find a matrix that represents a six-fold symmetry about the x_3 axis.

3

Review of Continuum Mechanics

Since piezoelectric materials exhibit coupling between their mechanical and electrical properties, in this chapter we review elementary principles from the foundations of *continuum mechanics* and linear elasticity. We present the definition of the stress vector $\boldsymbol{\tau}$ in a continuum, the corresponding definition of the stress tensor $\boldsymbol{T} := T_{ij}\boldsymbol{g}_i \otimes \boldsymbol{g}_j$, Cauchy's formula that relates $\boldsymbol{\tau}$ and $\boldsymbol{T}$, and the equilibrium equations in Section 3.1. The definition of the linear strain tensor $\boldsymbol{S} := S_{ij}\boldsymbol{g}_i \otimes \boldsymbol{g}_j$ is given in Section 3.2. The definition of the strain energy density, as well as some example calculations of strain energy for specific common structural elements, is the topic of Section 3.3. Generalized Hooke's law is presented in Section 3.4, which specifies the constitutive laws that are employed in linear elasticity. Finally, a summary of the initial-boundary value problem that underlies the formulation of mechanics for linearly elastic materials is presented in Section 3.5.

3.1 Stress

Stress and strain are the field variables that are used to relate the forces that are applied to a material continuum to the resulting deformation. Consider the continuum of material shown in Figure 3.1 that contains the volume Ω with surface $\partial\Omega$.

When the body is subjected to external loads and deformation occurs, the portion of the body that is outside Ω exerts forces on Ω, and vice versa. The stress vector $\boldsymbol{\tau}$ acting on Ω at a point $\boldsymbol{x}$ on the surface $\partial\Omega$ is defined as the limit

$$\boldsymbol{\tau} = \lim_{\Delta A \to 0} \frac{\Delta \boldsymbol{f}}{\Delta A},$$

where $\Delta \boldsymbol{f}$ is the force acting on the surface patch $\Delta A \subset \partial\Omega$. Stress has the units of force per area, or $\mathrm{N/m^2}$ in SI units. A Pascal $\mathrm{Pa} = 1\mathrm{N/m^2}$, and stresses are often expressed in applications in terms of $\mathrm{MPa} = 1 \times 10^6\,\mathrm{Pa}$.

Vibrations of Linear Piezostructures, First Edition. Andrew J. Kurdila and Pablo A. Tarazaga.

This Work is a co-publication between John Wiley & Sons Ltd and ASME Press.

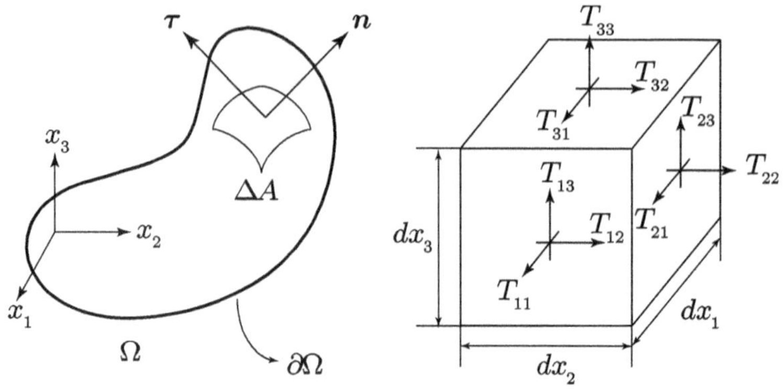

Figure 3.1 (Left) Continuum body Ω having surface $\partial\Omega$ and the stress vector τ acting on surface ΔA having unit normal $\boldsymbol{n}$, (Right) Stress vectors acting on the faces of a parallelepiped aligned with the coordinate directions

3.1.1 The Stress Tensor

Figure 3.1 (Right) depicts the case when the domain Ω is selected to be a volume in the shape of a differential cube that is aligned with the Cartesian coordinate axes and has side lengths dx_1, dx_2, and dx_3. The components T_{ij} of the stress vectors that act on each face are shown in the figure. If a face has an outward normal along the ith positive coordinate direction, T_{ij} denotes the jth component of the stress vector acting on the face. On the other hand, if the face has an outward normal direction along the ith negative coordinate direction, T_{ij} is the negative of the jth component of the stress vector. For example, the stress vectors acting on the right and left faces of the infinitesimally small cube are given, respectively, by

$$\tau_r := \begin{Bmatrix} T_{21} \\ T_{22} \\ T_{23} \end{Bmatrix}, \qquad \tau_l := \begin{Bmatrix} -T_{21} \\ -T_{22} \\ -T_{23} \end{Bmatrix}.$$

It can be shown that the nine entries T_{ij}, where $i, j = 1, 2, 3$, define the components of a second order tensor, the stress tensor, relative to bases along the coordinate axes x_1, x_2, x_3. It is convenient to organize the components T_{ij} in matrix form

$$\boldsymbol{T} := \begin{bmatrix} T_{11} & T_{12} & T_{13} \\ T_{21} & T_{22} & T_{23} \\ T_{31} & T_{32} & T_{33} \end{bmatrix}.$$

Under the common assumption [15] that there are no body moments proportional to volume, a moment summation shows that the stress tensor is symmetric, so that $T_{ij} = T_{ji}$ for $i, j = 1, 2, 3$. Since $\boldsymbol{T}$ is a second order tensor, its components T_{ij} transform under a rotation of basis r_{ij} to new components T'_{kl} via the transformation law

$$T'_{k\ell} = r_{ki} r_{\ell j} T_{ij}.$$

3.1.2 Cauchy's Formula

We often must determine the stress vector at a point that lies on some surface that is arbitrarily oriented, one that is not aligned with the coordinate axes. Cauchy's formula gives an expression for the stress vector $\boldsymbol{\tau}$ at a point that lies on a surface having unit normal $\boldsymbol{n}$. Cauchy's formula can be written in any of the equivalent forms $\boldsymbol{\tau} = \boldsymbol{T}\boldsymbol{n}$, $\tau_i = T_{ij} n_j$, or

$$\begin{Bmatrix} \tau_1 \\ \tau_2 \\ \tau_3 \end{Bmatrix} = \begin{bmatrix} T_{11} & T_{12} & T_{13} \\ T_{21} & T_{22} & T_{23} \\ T_{31} & T_{32} & T_{33} \end{bmatrix} \begin{Bmatrix} n_1 \\ n_2 \\ n_3 \end{Bmatrix}. \tag{3.1}$$

The derivation of Cauchy's formula is carried out by considering the equilibrium of a prototypical volume that has the shape of a tetrahedron as shown in Figure 3.2. If the beveled face in the figure has area da and surface normal $\boldsymbol{n}$, the areas of the faces of the tetrahedron that are normal to the x_1, x_2, x_3 axes are, respectively, $n_1 da, n_2 da, n_3 da$. If T_{ij} is the value of the stress tensor at the origin, the stress on the face that is perpendicular to the x_1 axis can be expanded in a Taylor series as

$$T_{11}(\boldsymbol{x} + \boldsymbol{\Delta}_1) = T_{11}(\boldsymbol{x}) + \frac{\partial T_{11}}{\partial x_2}\frac{dx_2}{3} + \frac{\partial T_{11}}{\partial x_3}\frac{dx_3}{3} + O(dx_i^2) \approx T_{11}(\boldsymbol{x}) + O(dx_i),$$

where $\boldsymbol{\Delta}_1 := \left[0, dx_2/3, dx_3/3\right]^T$ is the offset from the origin to the point of application of T_{11}. Each Δ_i is of the order $O(dx_i)$ for $i = 1, 2, 3$. Equilibrium in the 1 direction, for example, yields the equation

$$\begin{aligned} &\rho_m \left(\frac{1}{2}\frac{1}{3} dx_1 dx_2 dx_3\right) \frac{\partial^2 u_1}{\partial t^2} \\ &\qquad = \tau_1 da - T_{11} n_1 da - T_{21} n_2 da - T_{31} n_3 da + O(dx_i) da, \end{aligned}$$

where dx_3 is the height of the tetrahedron, ρ_m is the mass density, and u_i is the displacement in the i-direction. If we divide by da and take the limit as $dh \to 0$, this equation reduces to either of the forms $\tau_1 = T_{1j} n_j$ or

$$\tau_1 = \begin{bmatrix} T_{11} & T_{21} & T_{31} \end{bmatrix} \begin{Bmatrix} n_1 \\ n_2 \\ n_3 \end{Bmatrix}.$$

When this argument is repeated by summing the forces in either the 2 or 3 directions, Equation 3.1 is obtained.

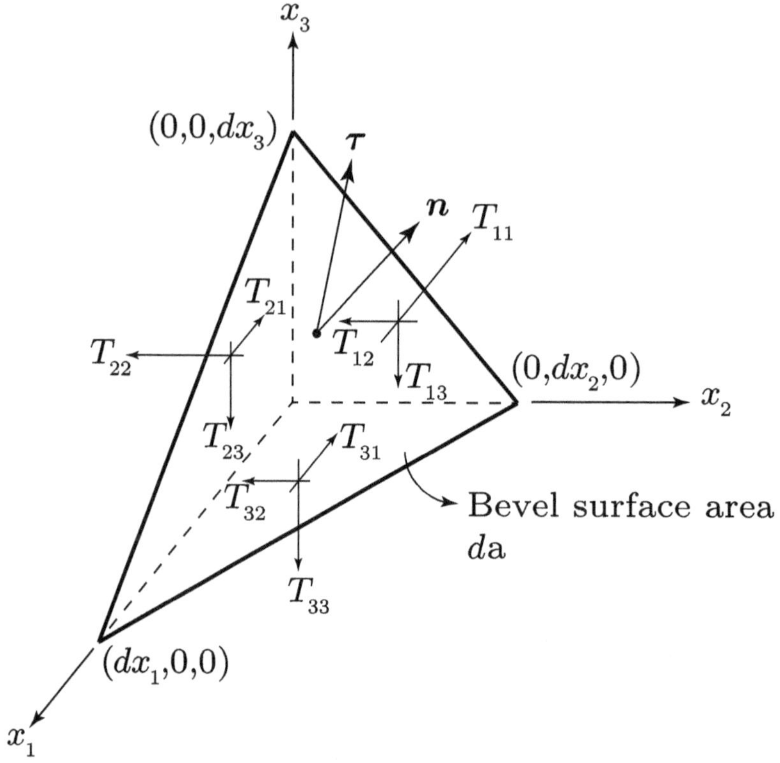

Figure 3.2 Tetrahedron with surface normal $\boldsymbol{n}$ used to derive Cauchy's formula

3.1.3 The Equations of Equilibrium

The differential volume in Figure 3.3 is centered at the point $\boldsymbol{x}$, and the stress tensor at $\boldsymbol{x}$ is again denoted $T_{ij}(\boldsymbol{x})$. By definition, the stress vector on each face of the cube is defined in terms of the stress tensor $T_{ij}(\boldsymbol{x} + \boldsymbol{\Delta}_i)$ where $\boldsymbol{\Delta}_i$ is a vector that locates the centroid of the face with a normal along the ith coordinate direction. For example $\boldsymbol{\Delta}_1 := \left[0, dx_2/3, dx_3/3\right]^T$, and similarly for Δ_2, Δ_3. Newton's law for the motion of the cube in the x_2 direction can be expressed as follows:

$$\begin{aligned}
\rho_m \frac{\partial^2 u_2}{\partial t^2} dx_1 dx_2 dx_3 &= T_{22}(\boldsymbol{x} + \boldsymbol{\Delta}_2) dx_1 dx_3 - T_{22}(\boldsymbol{x} - \boldsymbol{\Delta}_2) dx_1 dx_3 \\
&\quad + T_{32}(\boldsymbol{x} + \boldsymbol{\Delta}_3) dx_1 dx_2 - T_{32}(\boldsymbol{x} - \boldsymbol{\Delta}_3) dx_1 dx_2 \\
&\quad + T_{12}(\boldsymbol{x} + \boldsymbol{\Delta}_1) dx_2 dx_3 - T_{12}(\boldsymbol{x} - \boldsymbol{\Delta}_1) dx_2 dx_3 \\
&\quad + f_{b2} dx_1 dx_2 dx_3
\end{aligned}$$

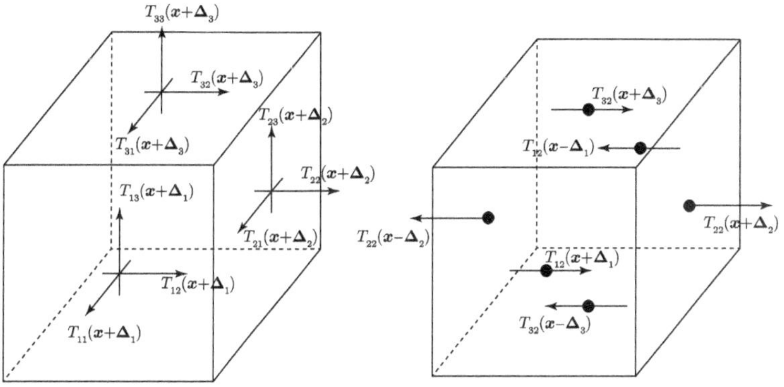

Figure 3.3 (Left) Differential cube with surface stresses, (Right) All stresses in x_1 direction

where f_{bi} is the body force per unit volume in the x_i direction for $i = 1, 2, 3$. We expand each stress tensor in terms of a Taylor series expansion, which yields

$$\begin{aligned}
&\rho_m \frac{\partial^2 u_2}{\partial t^2} dx_1 dx_2 dx_3 \\
&\quad = \left(T_{22} + \frac{\partial T_{22}}{\partial x_2}\frac{dx_2}{2} - \left(T_{22} - \frac{\partial T_{22}}{\partial x_2}\frac{dx_2}{2} \right)\right) dx_1 dx_3 + O(dx_1 dx_2^2 dx_3) \\
&\quad + \left(T_{32} + \frac{\partial T_{32}}{\partial x_3}\frac{dx_3}{2} - \left(T_{32} - \frac{\partial T_{32}}{\partial x_3}\frac{dx_3}{2} \right)\right) dx_1 dx_2 + O(dx_1 dx_2 dx_3^2) \\
&\quad + \left(T_{12} + \frac{\partial T_{12}}{\partial x_1}\frac{dx_1}{2} - \left(T_{12} - \frac{\partial T_{12}}{\partial x_1}\frac{dx_1}{2} \right)\right) dx_2 dx_3 + O(dx_1^2 dx_2 dx_3) \\
&\quad + f_{b2} dx_1 dx_2 dx_3.
\end{aligned}$$

The final equation governing motion in the x_2 direction is obtained when we divide by $dx_1 dx_2 dx_3$ and take the limit as the volume of the differential element approaches zero. We conclude that

$$\rho_m \frac{\partial^2 u_2}{\partial t^2} = \frac{\partial T_{12}}{\partial x_1} + \frac{\partial T_{22}}{\partial x_2} + \frac{\partial T_{32}}{\partial x_3} + f_{b2}.$$

Similar equations can be derived by considering the equilibrium in the x_1 and x_3 directions. The equilibrium equations in all three directions are summarized by the single equation

$$\rho_m \frac{\partial^2 u_i}{\partial t^2} = \frac{\partial T_{ji}}{\partial x_j} + f_{bi}.$$

3.2 Displacement and Strain

Measures of strain are used to describe the deformation of material continua, and several types have been introduced that are suitable for different applications or problems. Two of the most common, *Green's strain tensor* and *Cauchy's strain tensor*, coincide in the case that deformation gradients are so small that nonlinear contribution can be neglected [15]. In this text we consider only the linear or infinitesimal strain tensor that appears in the equations of linear piezoelectricity.

Consider the body Ω depicted in Figure 3.4 that is deformed into a new configuration $\tilde{\Omega}$. Each point $\boldsymbol{x} \in \Omega$ is mapped into a new spatial location $\tilde{\boldsymbol{x}} \in \tilde{\Omega}$ during the deformation process. The displacement at each point $\boldsymbol{x} \in \Omega$ is defined as $\boldsymbol{u}(\boldsymbol{x}) = \tilde{\boldsymbol{x}} - \boldsymbol{x}$, so that $u_i(\boldsymbol{x}) = \tilde{x}_i - x_i$. The components of the infinitesimal or linear strain tensor S_{ij} are defined as

$$S_{ij} := \frac{1}{2}\left(\frac{\partial u_i}{\partial x_j} + \frac{\partial u_j}{\partial x_i}\right), \tag{3.2}$$

where u_1, u_2, u_3 are the displacements at point $\boldsymbol{x}$ along the x_1, x_2, x_3 directions, respectively. The units of strain are m/m. In common engineering applications the strain is often given in units of percent strain which is equal to 0.01 m/m, or in terms of microstrain that is equal to 10^{-6} m/m. It can be shown that the linear infinitesimal strain S_{ij} determines a second order tensor

$$\boldsymbol{S} := S_{ij}\boldsymbol{g}_i \otimes \boldsymbol{g}_j.$$

In the usual way, if $\{\boldsymbol{g}_i\}_{i=1,2,3}$ and $\{\boldsymbol{g}'_i\}_{i=1,2,3}$ are two bases, the components S_{ij} and S'_{ij} relative to these bases, respectively, satisfy the transformation law

$$S'_{ij} = r_{im} r_{jn} S_{mn}.$$

The definition of the linear strain tensor can be used to formulate small deformation mechanics of general three dimensional bodies. For example,

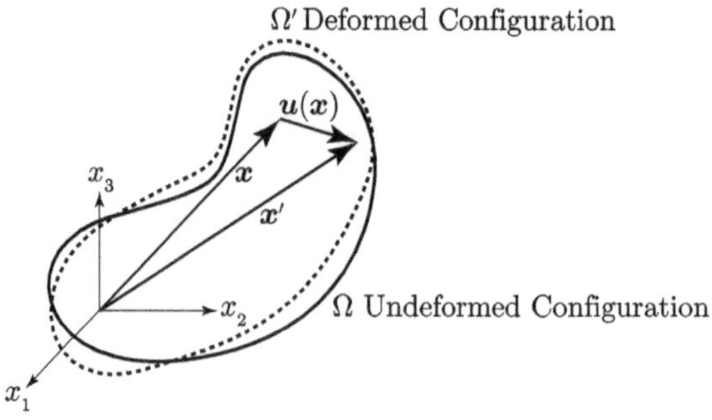

Figure 3.4 Undeformed configuration Ω, deformed configuration $\tilde{\Omega}$, and the deformation field $\boldsymbol{u}$

finite element formulations of linear elasticity or linear piezoelectricity of three dimensional continua employ the linear strain tensor. In contrast, for the simplest structures, we often derive the form of the strain components directly from approximations of the kinematics of a problem at hand. It is important to note that when we construct these specialized estimates of the strain S_{ij} that are adapted to the geometry of a problem at hand, the approximations may not coincide with the general definition of the strain given in Equation 3.2. In the following examples, we discuss such approximations for three such prototypical structures, the axial rod, the Bernoulli–Euler beam, and the Kirchoff–Love plate. The contrast between these specialized approximations and the general form for S_{ij} is evident when we compare the approximation in Equation 3.3 for the axial strain in a Bernoulli–Euler beam with Equation 3.2.

Example 3.2.1 In later examples in this text, active structures will be studied that are based on simple, iconic structures. Consider the column, or axial structural element, shown in Figure 3.5. This figure depicts the prototypical structural element known as an axial rod, which can be used as a starting point to formulate a model for piezoelectric stack actuators. In this structural element we assume that each plane parallel to the $x_1 - x_2$ coordinate plane remains plane during deformation and displaces uniformly in the x_3 direction. The kinematics of the axial rod are simple: it is assumed that the axial displacement u_3 is the only non-negligible displacement. We assume [11] that the displacements have the functional form

$$\begin{aligned} u_1 &= u_1(x_1, x_2, x_3) \approx 0, \\ u_2 &= u_2(x_1, x_2, x_3) \approx 0, \\ u_3 &= u_3(x_3), \end{aligned}$$

and consequently that the only non-negligible strain is given by

$$S_{33} = \frac{\partial u_3}{\partial x_3}.$$

It is important to observe that the transverse displacements u_1, u_2 are not generally equal to zero. If the lateral sides of the specimen are free, the displacements u_1 and u_2 may be nonzero due to Poisson's effect. In modeling an axial structural element, these effects are either assumed to be negligible or are assumed not to be of interest in the problem at hand. The simple functional form of the strain displacement relationship can be a source of insight into the nature of axial strains. If, for example, the displacement is a linear function of the coordinate x_3 and the base is fixed, the displacement can be

(Continued)

Example 3.2.1 (Continued)

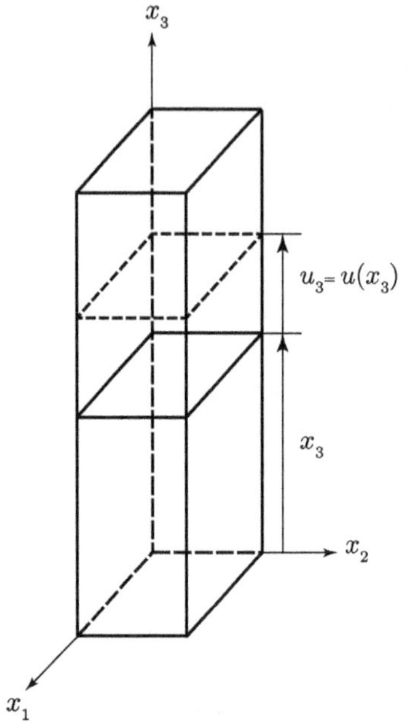

Figure 3.5 Axial rod geometry, coordinate alignment, and displacement

written in the form

$$u_3(x_3) = \frac{\Delta}{L} x_3,$$

where L is the length of the axial rod and Δ is the displacement of the end of the rod at $x_3 = L$. A direct calculation shows that the strain in this case is

$$S_{33} = \frac{\Delta}{L},$$

which agrees with our intuition and illustrates that the strain in one dimension can be interpreted approximately as the change in length divided by the original length.

Example 3.2.2 One of the most common architectures for building active structures uses piezoelectric material patches that are bonded to flexible beams. In this example we review the kinematics of a Bernoulli–Euler beam and derive a common approximation for its axial strain. Figure 3.6 illustrates

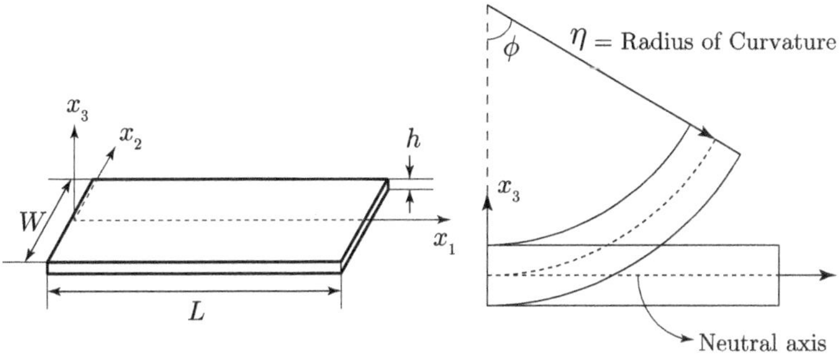

Figure 3.6 Beam geometry, coordinate alignment, and displacement

a Bernoulli–Euler beam with the origin of the x_1, x_2, x_3 frame at the base of the beam and the x_1 direction along the length of the beam. The length of the beam is L, the thickness of the beam is h, the width is W. Two assumptions dictate the kinematics of a Bernoulli–Euler beam. The first assumption stipulates that when the beam deflects it takes the form of a circular arc. η is the radius of curvature, and ϕ is the angle that measures the sector of the circular arc into which the beam deforms. The neutral axis, which is a fiber along the x_1 direction that does not change length, deforms into a circular arc having radius of curvature η. The second assumption requires that when the beam deforms, cross sections that were initially perpendicular to the neutral axis remain so after deformation. We approximate the strain in the x_1 direction of a typical fiber parallel to the neutral axis by calculating the ratio of its change in length to its original length. As shown in the figure, a fiber located at a distance x_3 from the neutral axis will have a length $(\eta - x_3)\phi$. The strain S_{11} is then approximated by the ratio

$$S_{11} \approx \frac{\Delta}{L} = \frac{(\eta - x_3)\phi - \eta\phi}{\eta\phi} = -\frac{x_3}{\eta}.$$

The radius of curvature η can be related to the displacement $u_3 = u_3(x_1)$ in the x_3 direction along the neutral axis by the formula

$$\frac{1}{\eta} = \frac{\frac{\partial^2 u_3}{\partial x_1^2}}{\sqrt{1 + \left(\frac{\partial u_3}{\partial x_1}\right)^2}} \approx \frac{\partial^2 u_3}{\partial x_1^2}.$$

(Continued)

Example 3.2.2 (Continued)

Combining these two equations gives the most common form of the strain-displacement relations for a Bernoulli–Euler beam,

$$S_{11} = -x_3 \frac{\partial^2 u_3}{\partial x_1^2}. \tag{3.3}$$

Example 3.2.3 As a final example of a prototypical structural component, we consider the strain displacement relationships in a thin plate. There are in fact a wide variety of models that have been derived for different classes of plates and shells. See [49] for some popular and classic representations, as well as nonlinear kinematic models for piezoelectric laminates. Reference [4] contains a good account of functional analytic details of a variety of piezoelectrically actuated plates. The only example of plates or shells considered in this text are those represented via the classical Kirchoff plate model. While the model is limited to linear response regimes, it is intuitively quite similar to the Bernoulli–Euler model for beams and makes an excellent starting point for more refined plate analyses.

By assumption, the thickness of a Kirchoff plate must be small in comparison to the local radius of curvature anywhere on the deformed plate. Kirchoff plate theory also assumes that the normal stresses in the direction perpendicular to the plate are negligible in comparison to the in-plane normal stresses. Finally, it is assumed that a fiber that is originally normal to the mid-plane or neutral surface of the plate remains perpendicular to the plate during deformation (Figure 3.7). With these assumptions, the strain-displacement relationships in the Kirchoff theory for a flat plate take the form

$$S_{11} = \frac{\partial u_1}{\partial x_1} - x_3 \frac{\partial^2 u_3}{\partial x_1^2},$$

$$S_{22} = \frac{\partial u_2}{\partial x_2} - x_3 \frac{\partial^2 u_3}{\partial x_2^2},$$

$$S_{12} = \frac{1}{2}\left(\frac{\partial u_2}{\partial x_1} + \frac{\partial u_1}{\partial x_2}\right) - x_3 \frac{\partial^2 u_3}{\partial x_1 \partial x_2}.$$

The in-plane normal strains S_{11} and S_{22} have similar form and can be understood intuitively by comparing these strains to the strain field in the Bernoulli–Euler beam in Equation 3.3. The normal strain S_{11} is given by superposition of two parts. The contribution $\frac{\partial u_1}{\partial x_1}$ is due to the compression or tension of the specimen along the axial x_1 direction. The contribution $-x_3 \frac{\partial^2 u_3}{\partial x_1^2}$ is due to plate bending in the $x_1 - x_3$ plane. This term is identical in structure to the representation of the strain in a Bernoulli–Euler beam in the $x_1 - x_3$ plane. It can be interpreted as the change in length that a fiber undergoes as the plate is deformed into a circular arc in the $x_1 - x_3$ plane. The derivation of this strain expression follows exactly the same arguments as used in our discussion of the Bernoulli–Euler beam. A similar interpretation holds for the two terms that constitute the strain S_{22}. The shear strain S_{12} is also the superposition of two terms. The contribution $\frac{1}{2}\left(\frac{\partial u_1}{\partial x_2} + \frac{\partial u_2}{\partial x_1}\right)$ is due to in-plane deformations. Again, the contribution $-x_3 \frac{\partial^2 u_3}{\partial x_1 \partial x_2}$ is due to bending in the $x_1 - x_3$ and $x_2 - x_3$ planes. The plate model serves as the foundation for approximations of composite piezoelectric plates studied in Sections 6.4 and Example 6.4.1.

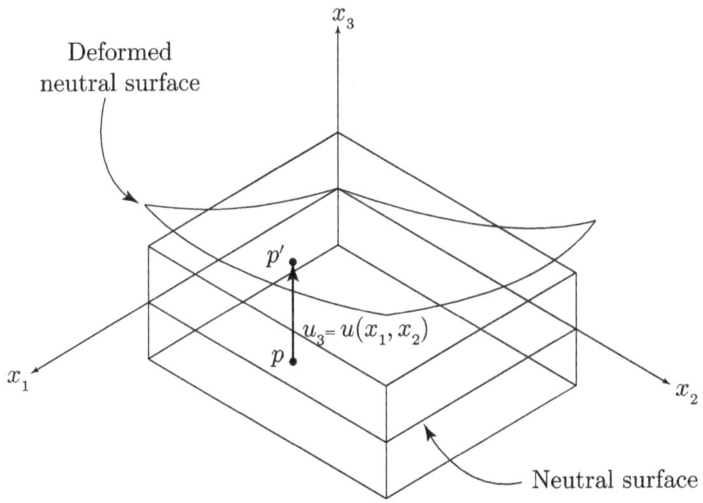

Figure 3.7 Geometry of the thin, rectangular, Kirchoff plate

3.3 Strain Energy

We will see in Chapter 5 that a thermodynamic analysis of linear piezoelectric materials provides a rich framework for describing their constitutive laws. The thermodynamic formalism relies on the study of energy functionals such as the internal energy density U and the electric enthalpy density H of a piezoelectric continua. In the case of linearly elastic materials, or for those in which the piezoelectric effect is negligible, it will be shown that both U and H reduce to the strain energy density $\mathcal{U}_0$. In the next section, we exploit properties of the strain energy density function to prove symmetry properties of linear constitutive laws for linearly elastic continua. In this section we define and briefly review the strain energy density of a linearly elastic continua [33].

The total stress power P in a linearly elastic continuum Ω is the rate of work done by the stresses T_{ij} acting through the strains S_{ij}

$$P = \int_\Omega T_{ij} \frac{dS_{ij}}{dt} dv,$$

integrated over the volume Ω. We say that $\mathcal{U}_0$ is a strain energy density function for the material continuum Ω if $\mathcal{U}_0$ is a continuously differentiable function of the strains S_{ij} and $P = \int_\Omega \frac{d\mathcal{U}_0}{dt} dv$. Since $P = \int_\Omega \frac{\partial \mathcal{U}_0}{\partial S_{ij}} \frac{dS_{ij}}{dt} dv = \int_\Omega T_{ij} \frac{dS_{ij}}{dt} dv$ for any subdomain Ω, it must be that case that

$$T_{ij} = \frac{\partial \mathcal{U}_0}{\partial S_{ij}} \tag{3.4}$$

whenever a strain energy density function exists. If a strain energy density function $\mathcal{U}_0$ exists, the material is said to be hyperelastic. For a linearly elastic material that has a quadratic strain energy density function, it is straightforward to show that there is a linear relationship among the stress and linear strain tensors. If we define

$$\mathcal{U}_0 := \frac{1}{2} C_{ijkl} S_{ij} S_{kl},$$

then a quick calculation shows that

$$\begin{aligned}\frac{\partial \mathcal{U}_0}{\partial S_{mn}} &= \frac{1}{2} C_{ijkl} \left(\frac{\partial S_{ij}}{\partial S_{mn}} S_{kl} + S_{ij} \frac{\partial S_{kl}}{\partial S_{mn}} \right) = \frac{1}{2} C_{ijkl} (\delta_{im} \delta_{jn} S_{kl} + S_{ij} \delta_{km} \delta_{ln}), \\ &= \frac{1}{2} (C_{mnkl} S_{kl} + C_{ijmn} S_{ij}) = C_{mnij} S_{ij} = T_{mn}.\end{aligned}$$

We have used the fact in the last line above that the symmetry condition $C_{ijkl} = C_{klij}$ follows from the definition of $\mathcal{U}_0$. The final result of this line of reasoning,

$$T_{mn} = C_{mnij} S_{ij},$$

is known as a generalized form of Hooke's Law. It is discussed more fully in the next section. As we will see in Section 5.3, the stress tensor in a linearly piezoelectric continua satisfies a similar identity, $T_{ij} = \frac{\partial \mathcal{U}}{\partial S_{ij}}$ where $\mathcal{U}$ is the internal energy. Equation 3.4 for the strain energy density $\mathcal{U}_0$ can be viewed as a special case of the corresponding identity for the internal energy $\mathcal{U}$.

3.4 Constitutive Laws for Linear Elastic Materials

Elasticity is the study of solid continua in which the deformation due to external mechanical loading is reversible, so that the deformed solid returns to its initial state when the loading is removed. Some of the simplest elastic materials are those that obey a generalized Hooke's law. We usually first encounter the one dimensional form of Hooke's law that states that the stress is proportional to the strain in a Hookean material, where Young's modulus Y is the constant of proportionality. The generalized Hooke's law states that the components of the stress tensor T_{ij} are a linear function of the components of the linear strain tensor S_{kl} and satisfy a formula having the form

$$T_{ij} = C_{ijkl}S_{kl}.$$

This equation should be interpreted as a relationship between the components of the second order stress tensor $\boldsymbol{T} := T_{ij}\mathbf{g}_i \otimes \mathbf{g}_j$, the second order strain tensor $\boldsymbol{S} := S_{kl}\mathbf{g}_k \otimes \mathbf{g}_l$, and a fourth order stiffness tensor of elastic constants $\boldsymbol{C} := C_{ijkl}\mathbf{g}_i \otimes \mathbf{g}_j \otimes \mathbf{g}_k \otimes \mathbf{g}_l$. Without consideration of symmetry, there are 81 coefficients that define the 4th order tensor C_{ijkl}. Fortunately, there are numerous symmetries that reduce the number of distinct values for the elastic constants. Since the stress tensor and strain tensors are symmetric, with $T_{ij} = T_{ji}$ and $S_{kl} = S_{lk}$, it must be that

$$C_{ijkl} = C_{jikl},$$
$$C_{ijkl} = C_{ijlk}.$$

Since there are 45 independent symmetry conditions appearing in these equations, we conclude that the symmetry of the stress and strain tensors imply that there are at most 36 independent elastic moduli in the components of the tensor C_{ijkl}.

However, more can be said for linearly elastic materials. The existence of a strain energy density function $\mathcal{U}_0$ for an elastic solid additionally implies that

$$C_{ijkl} = C_{klij}.$$

We must recall some properties of the strain energy density function $\mathcal{U}_0$ to see why this additional symmetry holds. By definition, the strain energy density function $\mathcal{U}_0$ provides a relationship between the entries of the stress tensor and strain tensor by requiring that

$$T_{ij} = \frac{\partial \mathcal{U}_0}{\partial S_{ij}}.$$

One implication of this identity is that we can change the order of differentiation of the strain energy density and conclude that

$$\frac{\partial T_{ij}}{\partial S_{kl}} = \frac{\partial^2 \mathcal{U}_0}{\partial S_{kl}\partial S_{ij}} = \frac{\partial^2 \mathcal{U}_0}{\partial S_{ij}\partial S_{kl}} = \frac{\partial T_{kl}}{\partial S_{ij}}.$$

We next combine these identities with the generalized Hooke's law in a pair of expressions

$$\frac{\partial T_{ij}}{\partial S_{kl}} = \frac{\partial}{\partial S_{kl}}(C_{ijmn}S_{mn}) = C_{ijmn}\delta_{km}\delta_{ln} = C_{ijkl},$$
$$\frac{\partial T_{kl}}{\partial S_{ij}} = \frac{\partial}{\partial S_{ij}}(C_{klmn}S_{mn}) = C_{klmn}\delta_{im}\delta_{jn} = C_{klij},$$

and the symmetry conditions $C_{ijkl} = C_{klij}$ follow.

All of these symmetry conditions imply that there are at most 21 independent coefficients in the stiffness tensor C_{ijkl}. It is common practice in many applications to rewrite the generalized Hooke's law in matrix form. The matrix form is constructed by renumbering the independent entries of the stress and strain tensors as single index 6-tuples. We define the 6×1 arrays $\boldsymbol{T}$ and $\boldsymbol{S}$ that contain the independent stress tensor and strain tensor components, respectively, as

$$\boldsymbol{T} := \begin{Bmatrix} T_1 \\ T_2 \\ T_3 \\ T_4 \\ T_5 \\ T_6 \end{Bmatrix} = \begin{Bmatrix} T_{11} \\ T_{22} \\ T_{33} \\ T_{23} \\ T_{13} \\ T_{12} \end{Bmatrix} \quad \text{and} \quad \boldsymbol{S} := \begin{Bmatrix} S_1 \\ S_2 \\ S_3 \\ S_4 \\ S_5 \\ S_6 \end{Bmatrix} = \begin{Bmatrix} S_{11} \\ S_{22} \\ S_{33} \\ 2S_{23} \\ 2S_{13} \\ 2S_{12} \end{Bmatrix}.$$

Carefully note that we define $S_4, S_5, S_6 \equiv 2S_{23}, 2S_{13}, 2S_{12}$ in the compact engineering notation. This choice enables one to interpret S_4, S_5, S_6 as corresponding to the angular change in axes that initially form right angles in the undeformed continua. See [15] for a detailed discussion. The matrix form of the generalized Hooke's law is now written in the explicit form

$$\begin{Bmatrix} T_1 \\ T_2 \\ T_3 \\ T_4 \\ T_5 \\ T_6 \end{Bmatrix} = \begin{bmatrix} C_{11} & C_{12} & \cdots & C_{16} \\ C_{21} & C_{22} & \cdots & C_{26} \\ \vdots & \vdots & \ddots & \vdots \\ C_{61} & C_{62} & \cdots & C_{66} \end{bmatrix} \begin{Bmatrix} S_1 \\ S_2 \\ S_3 \\ S_4 \\ S_5 \\ S_6 \end{Bmatrix},$$

or via the summation convention as

$$T_i = C_{ij}S_j.$$

The coefficients $C_{ij} = C_{ji}$ are symmetric, and there are 21 independent constants in general in the compact matrix notation.

Additional constraints on the number of independent entries in the constitutive tensor C_{ijkl}, or equivalently the constitutive matrix C_{ij}, can be deduced by defining further symmetry conditions on the properties of the material. This is an important topic for piezoelectric materials where the coupling of electrical and mechanical properties depends on some type of asymmetry of the underlying crystalline structure. Recall that in Section 2.3 we discussed various crystal structures and observed that each unit cell has an associated set of symmetry transformations. The additional constraints due to material asymmetry are constructed by enforcing invariance of the constitutive laws under symmetry transformations. When the general form of the transformation law for the fourth order stiffness tensor is written as $C'_{ijkl} = r_{im}r_{jn}r_{ko}r_{lp}C_{mnop}$, the condition that the constitutive law is invariant under the change of coordinates described by r_{im} is written as

$$C_{ijkl} = r_{im}r_{jn}r_{ko}r_{lp}C_{mnop}.$$

The detailed inventory of all the symmetry transformations for each specific unit cell that arises in crystallography requires a level of detail that goes beyond our goals in this text. Again, the interested reader can refer to [43, 44] for a discussion of crystal structure and symmetries for some common piezoelectric materials, or to [13] for a description of symmetry operations associated with all the crystal classes.

There is a potential source of confusion when reading background material from continuum mechanics or elasticity that the reader should keep in mind. It is common practice in some accounts of continuum mechanics or linear elasticity to define general types of materials that share symmetry properties, and these collections can include many similar, but distinct, crystal classes. To make matters more confusing, some of the names for these categories of materials are identical to the names of particular crystal classes. These categories are referred to as monoclinic materials, triclinic materials, orthotropic materials, or transversely isotropic materials, for instance. These more general, macroscopic symmetry categories include materials whose unit cells share a common set of underying symmetry operations, but they are not crystal classes. For example, by consulting the Table A.2 in [13], we find that crystal classes with International Symbol (4,5:monoclinic), (8:orthorhombic), (11,15: tetragonal), (22, 23, 26, 27:hexagonal), and (29, 32:cubic) all have a plane of material symmetry. In other words, if we choose the $x_1 - x_2$ plane as the plane of symmetry, they all share the symmetry transformation associated with reflection about a plane perpendicular to x_3. They have unit cells that are invariant under the reflection transformation r_{ij} that has the form

$$r_{ij} = \begin{bmatrix} 1 & 0 & 0 \\ 0 & 1 & 0 \\ 0 & 0 & -1 \end{bmatrix}.$$

All of these crystal classes can be viewed as belonging to the general category of "monoclinic materials", where monoclinic materials are defined in continuum mechanics texts to be those materials that have a plane of symmetry. To be sure, some of the crystal classes mentioned above have many additional symmetries, but when we refer to "monoclinic materials" in continuum mechanics, it is to this larger category that we often refer. We briefly summarize some of the more common material symmetry categories and discuss the form of their constitutive laws.

3.4.1 Triclinic Materials

Triclinic linearly elastic materials are the most general in that they do not have any associated symmetry transformations, other than the identity. Generalized Hooke's law for triclinic materials has the form

$$\begin{Bmatrix} T_1 \\ T_2 \\ T_3 \\ T_4 \\ T_5 \\ T_6 \end{Bmatrix} = \begin{bmatrix} C_{11} & C_{12} & \cdots & C_{16} \\ C_{12} & C_{22} & \cdots & C_{26} \\ \vdots & \vdots & \ddots & \vdots \\ C_{16} & C_{26} & \cdots & C_{66} \end{bmatrix} \begin{Bmatrix} S_1 \\ S_2 \\ S_3 \\ S_4 \\ S_5 \\ S_6 \end{Bmatrix},$$

where all 21 coefficients are in general nonzero and independent.

3.4.2 Monoclinic Materials

Monoclinic materials are those that have a plane of material symmetry, so that the constitutive laws are invariant with respect to reflection about the plane of symmetry. If we arrange the coordinate axes so that the x_3 axis is perpendicular to the plane of symmetry (x_1, x_2), generalized Hooke's law for monoclinic materials has the structure

$$\begin{Bmatrix} T_1 \\ T_2 \\ T_3 \\ T_4 \\ T_5 \\ T_6 \end{Bmatrix} = \begin{bmatrix} C_{11} & C_{12} & C_{13} & 0 & 0 & C_{16} \\ C_{12} & C_{22} & C_{23} & 0 & 0 & C_{26} \\ C_{13} & C_{23} & C_{33} & 0 & 0 & C_{36} \\ 0 & 0 & 0 & C_{44} & C_{45} & 0 \\ 0 & 0 & 0 & C_{45} & C_{55} & 0 \\ C_{16} & C_{26} & C_{36} & 0 & 0 & C_{66} \end{bmatrix} \begin{Bmatrix} S_1 \\ S_2 \\ S_3 \\ S_4 \\ S_5 \\ S_6 \end{Bmatrix}. \tag{3.5}$$

Monoclinic materials in general have 13 independent coefficients in C_{ij}.

3.4.3 Orthotropic Materials

Orthotropic materials are those that have three orthogonal planes of material symmetry. Their constitutive laws are invariant with respect to reflection perpendicular to the three planes of symmetry. When the three coordinates x_1, x_2, x_3 are defined to be perpendicular to the planes of material symmetry, generalized

Hooke's law is expressed as

$$\begin{Bmatrix} T_1 \\ T_2 \\ T_3 \\ T_4 \\ T_5 \\ T_6 \end{Bmatrix} = \begin{bmatrix} C_{11} & C_{12} & C_{13} & 0 & 0 & 0 \\ C_{12} & C_{22} & C_{23} & 0 & 0 & 0 \\ C_{13} & C_{23} & C_{33} & 0 & 0 & 0 \\ 0 & 0 & 0 & C_{44} & 0 & 0 \\ 0 & 0 & 0 & 0 & C_{55} & 0 \\ 0 & 0 & 0 & 0 & 0 & C_{66} \end{bmatrix} \begin{Bmatrix} S_1 \\ S_2 \\ S_3 \\ S_4 \\ S_5 \\ S_6 \end{Bmatrix}. \tag{3.6}$$

Orthotropic materials in general have 9 independent coefficients in C_{ij}.

3.4.4 Transversely Isotropic Materials

Transversely isotropic materials are also of specific interest when modeling certain types of piezoelectric materials. These materials have three mutually perpendicular planes of symmetry, and they additionally exhibit rotational symmetry about one axis that is perpendicular to one of the planes of symmetry. If, for example, the x_1, x_2, x_3 coordinate axes are selected to be perpendicular to the planes of material symmetry, and in addition the x_3 axis is the axis of rotational symmetry, then generalized Hooke's law can be written as

$$\begin{Bmatrix} T_1 \\ T_2 \\ T_3 \\ T_4 \\ T_5 \\ T_6 \end{Bmatrix} = \begin{bmatrix} C_{11} & C_{12} & C_{13} & 0 & 0 & 0 \\ C_{12} & C_{11} & C_{13} & 0 & 0 & 0 \\ C_{13} & C_{13} & C_{33} & 0 & 0 & 0 \\ 0 & 0 & 0 & C_{44} & 0 & 0 \\ 0 & 0 & 0 & 0 & C_{44} & 0 \\ 0 & 0 & 0 & 0 & 0 & \frac{1}{2}(C_{11} - C_{12}) \end{bmatrix} \begin{Bmatrix} S_1 \\ S_2 \\ S_3 \\ S_4 \\ S_5 \\ S_6 \end{Bmatrix}. \tag{3.7}$$

The matrix C_{ij} contains at most 5 independent entries in transversely isotropic materials.

3.5 The Initial-Boundary Value Problem of Linear Elasticity

In this section we describe the initial-boundary value problem that governs a linearly elastic continuum. Figure 3.8 illustrates the linearly elastic body Ω that is subject to kinematic and stress boundary conditions. The surface $\partial\Omega$ of the body Ω is partitioned into two disjoint surfaces, the traction boundary $\partial\Omega_T$ and the displacement boundary $\partial\Omega_u$. This decomposition covers the surface $\partial\Omega$.

$$\begin{aligned} \partial\Omega &= \partial\Omega_u \cup \partial\Omega_T, \\ \emptyset &= \partial\Omega_u \cap \partial\Omega_T. \end{aligned}$$

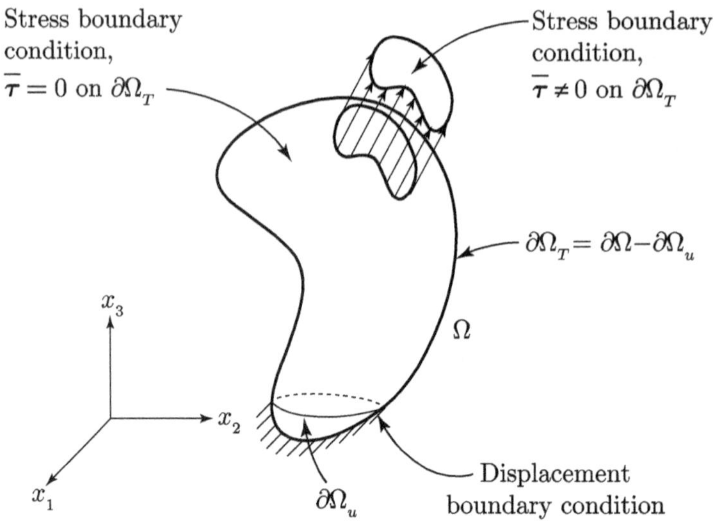

Figure 3.8 Linearly elastic body Ω, applied external stress field $\overline{\tau}$, boundary $\partial\Omega$, traction boundary $\partial\Omega_T$, and displacement boundary $\partial\Omega_u$

The traction boundary $\partial\Omega_T$ defines the portion of the boundary on which traction boundary conditions are imposed, while the displacement boundary $\partial\Omega_u$ specifies the portion of the boundary on which kinematic boundary conditions are imposed. The equations of linear elasticity are comprised of the equilibrium equations discussed in Section 3.1.3, the strain-displacement equations derived in Section 3.2, and the constitutive laws presented in Equations 3.4. We seek a displacement field $u_i = u_i(t, \boldsymbol{x})$, stress tensor $T_{ij} = T_{ij}(t, \boldsymbol{x})$ and linear strain tensor $S_{ij} = S_{ij}(t, \boldsymbol{x})$ that satisfy

$$\rho_m \frac{\partial^2 u_i}{\partial t^2} = \frac{\partial T_{ji}}{\partial x_j} + f_{bi}, \qquad \text{Equilibrium equations,}$$

$$S_{ij} = \frac{1}{2}\left(\frac{\partial u_i}{\partial x_j} + \frac{\partial u_j}{\partial x_i}\right), \qquad \text{Strain/displacement relationships,}$$

$$T_{ij} = C_{ijkl} S_{kl}, \qquad \text{Constitutive Law,}$$

for $(t, \boldsymbol{x}) \in \mathbb{R}^+ \times \Omega$, subject to the boundary conditions

$$T_{ij} n_j = \overline{\tau}_i, \qquad \text{for } (t, \boldsymbol{x}) \in \mathbb{R}^+ \times \partial\Omega_T,$$
$$u_i = \overline{u}_i, \qquad \text{for } (t, \boldsymbol{x}) \in \mathbb{R}^+ \times \partial\Omega_u,$$

and subject to the initial conditions

$$u_i = \overline{u}_{i,0}, \quad \text{for } \boldsymbol{x} \in \Omega,$$
$$\frac{\partial u_i}{\partial t} = \overline{v}_{i,0}, \quad \text{for } \boldsymbol{x} \in \Omega,$$

when $t = 0$. In these equations $\overline{\boldsymbol{\tau}}$ is the applied surface stress vector acting on $\partial\Omega_T$, $\overline{\boldsymbol{u}}$ is the displacement prescribed on $\partial\Omega_u$, $\overline{\boldsymbol{u}}_0$ is the initial displacement field defined on Ω, and $\overline{\boldsymbol{v}}_0$ is the initial velocity field defined on Ω.

Classical texts that study continuum mechanics and linear elasticity have derived analytical solutions of these equations under various assumptions on the geometry of the domain, the nature of the external loading, and the type of boundary and initial conditions. Even though the equations of linear elasticity can be reduced to a set of linear partial differential equations in terms of the displacements alone, analytic solutions are not available for general domains. Numerical approximations, and perhaps most significantly finite element or finite difference methods, have been studied in great detail over the years for solutions of the initial-boundary value problem over general domains Ω. [19, 51]

3.6 Problems

Problems 3.6.1 Derive Cauchy's formula

$$\tau_i = T_{ij} n_j$$

for $i = 2, 3$.

Problems 3.6.2 Derive the equations of equilibrium

$$\rho_m \frac{\partial^2 u_i}{\partial t^2} = \frac{\partial T_{ji}}{\partial x_j} + f_{bi}$$

for $i = 1, 3$.

Problems 3.6.3 Use Cauchy's formula to show that the components T_{ij} of the stress tensor obey the transformation laws of a second order tensor.

Problems 3.6.4 Use the strain–displacement relationships to show that the components S_{ij} of the linear strain tensor obey the transformation laws of a second order tensor.

Problems 3.6.5 Suppose that the components of a stress tensor relative to the $\mathbf{g}_1, \mathbf{g}_2, \mathbf{g}_3$ axes are given as

$$T_{ij} = \begin{bmatrix} 2 & 0 & 0 \\ 0 & 3 & 0 \\ 0 & 0 & 4 \end{bmatrix} \quad \text{MPa}$$

Define a new basis system $\mathbf{g}_1', \mathbf{g}_2', \mathbf{g}_3'$ by rotating 30° about the $\mathbf{g}_2 = \mathbf{g}_2'$ axis. What are the components of the stress tensor relative to the new basis?

Problems 3.6.6 If the components of a stress tensor are given as in Problem 3.6 at the origin $(x_1, x_2, x_3) = \mathbf{0}$, what is the stress vector at the same location acting on a surface that has the equation

$$7x_1 + 2x_2 - 4x_3 = 0?$$

Note that this is the equation of a plane that passes through the origin.

Problems 3.6.7 Show that the constitutive laws for a monoclinic material have the form shown in Equation 3.5.

Problems 3.6.8 Show that the constitutive laws for an orthotropic materials have the form shown in Equation 3.6.

Problems 3.6.9 Show that the constitutive laws for a transversely isotropic materials have the form shown in Equation 3.7.

4

Review of Continuum Electrodynamics

To understand the coupling between the electrical and mechanical properties of piezoelectric continua, we must supplement the discussion of continuum mechanics and linear elasticity presented in Chapter 3 with the review of continuum electrodynamics in this chapter. The fundamental definitions of charge, current, charge density, current density, bound charge density, and free charge density are discussed in Section 4.1. The electric and magnetic fields are introduced in Section 4.2. The dynamic coupling of the electric and magnetic fields is determined by Maxwell's equations, which are introduced in Section 4.3. The polarization and electric displacement are discussed in Section 4.3.1, while the magnetization and magnetic field intensity are introduced in Section 4.3.2.

4.1 Charge and Current

In this book an ideal point charge has a magnitude of charge q and is located at some physical location $\boldsymbol{x}$ in Euclidean space $\mathbb{R}^3$. The units of charge are Coulombs C in SI units. The current i carried in a wire is the time rate of change dq/dt of the charge q and is given in SI units as Amps $A = C/s$. When considering a continua Ω, the total charge density ρ_e describes the distribution of charge per unit volume in terms of Coulombs per volume, or C/m^3. While we usually think of the current as flow of charge that is concentrated in a one dimensional domain such as an ideal wire, we can generalize this definition appropriately for continua in three dimensions. The current density $\boldsymbol{j}(\boldsymbol{x})$ is the current per unit volume in a three dimensional continua Ω at the point $\boldsymbol{x}$. The current density $\boldsymbol{j}$ is a vector in three dimensions and has units of A/m^3.

These definitions are used to categorize materials into conductors and insulators. We can be more precise about how conductors and insulators differ by decomposing the total electric charge density ρ_e into the free charge density $\rho_{e,f}$

Vibrations of Linear Piezostructures, First Edition. Andrew J. Kurdila and Pablo A. Tarazaga.

This Work is a co-publication between John Wiley & Sons Ltd and ASME Press.

and bound charge density $\rho_{e,b}$,

$$\rho_e := \rho_{e,f} + \rho_{e,b}.$$

The free charge density is also known as the mobile charge, and the bound charge is sometimes referred to as the polarization charge. All real materials conduct electricity to some degree. Still, we define insulators as materials in which the mobile charge $\rho_{e,f} \approx 0$, while in conductors $\rho_{e,f} \not\approx 0$.

4.2 The Electric and Magnetic Fields

While the stress $\boldsymbol{T}$ and strain $\boldsymbol{S}$ are perhaps the most critical field variables needed to understand continuum mechanics and linear elasticity as summarized in Chapter 3, the electric field $\boldsymbol{E}$ and magnetic field $\boldsymbol{B}$ are equally pivotal in understanding continuum electrodynamics. A simple physical interpretation of the electric field $\boldsymbol{E}$ is obtained by studying the force among finite collections of charged particles. Similarly, the magnetic field $\boldsymbol{B}$ is readily defined by determining the forces that act between wires that carry current. In both cases, a general definition of the field is achieved by passing to the limit as the charge or current is viewed as distributed over a continuum.

4.2.1 The Definition of the Static Electric Field

Figure 4.1 depicts a pair of stationary point charges $q(\boldsymbol{x})$ and $q(\xi)$ that are located at the points $\boldsymbol{x}$ and ξ, respectively. By Coulomb's law the magnitude of the repulsive force $\boldsymbol{F}$ acting between the two charges is inversely proportional to the squared distance between them and can be written as

$$\|\boldsymbol{F}\| = \frac{1}{4\pi\varepsilon_0} \frac{q(\boldsymbol{x})q(\xi)}{\|\boldsymbol{x} - \xi\|^2},$$

where $\varepsilon_0 := 8.854 \times 10^{-12}$F/m is the permittivity of free space. We define the static electric field $\boldsymbol{E}(\boldsymbol{x})$ at $\boldsymbol{x}$ due to the fixed point charge $q(\xi)$ at ξ to be the force that a unit charge at $\boldsymbol{x}$ would feel due to $q(\xi)$. In other words, the static electric field $\boldsymbol{E}(\boldsymbol{x})$ at $\boldsymbol{x}$ due to the point charge $q(\xi)$ is given by

$$\boldsymbol{E}(\boldsymbol{x}) := \frac{\boldsymbol{F}}{q(\boldsymbol{x})} = \frac{1}{4\pi\varepsilon_0} q(\xi) \frac{\boldsymbol{x} - \xi}{\|\boldsymbol{x} - \xi\|^3}.$$

The electric field has units of force per charge, or N/C = V/m in SI units. The static electric field induced by a finite number of point charges $\{q(\xi_k)\}_{k=1,\dots,n}$ is obtained by superposition. We have

$$\boldsymbol{E}(\boldsymbol{x}) := \frac{1}{4\pi\varepsilon_0} \sum_{k=1}^{n} q(\xi_k) \frac{\boldsymbol{x} - \xi_k}{\|\boldsymbol{x} - \xi_k\|^3}.$$

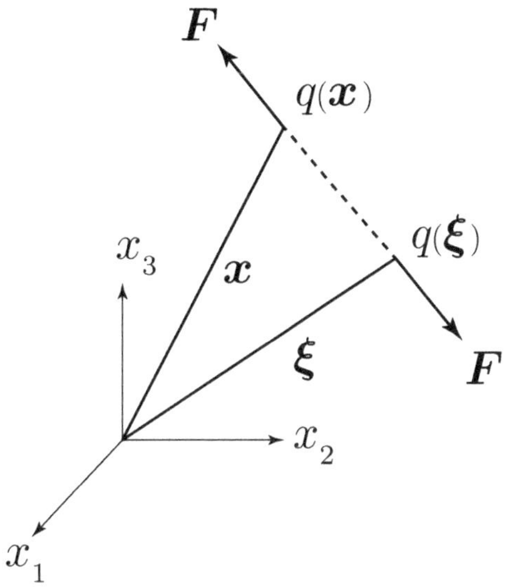

Figure 4.1 Point charges q_x, q_ξ located at points $\boldsymbol{x}, \xi$, respectively, position vectors $\boldsymbol{x}$ and ξ, and the repulsive force $\boldsymbol{F}$.

Finally, the static electric field for a continua Ω is obtained by taking the limit of this expression as the number of points become infinite and the summation passes to an integral. We introduce the total charge density ρ_e in the continuum Ω and replace the finite sum by an integral over the volume Ω.

$$\boldsymbol{E}(\boldsymbol{x}) := \frac{1}{4\pi\varepsilon_0}\int_\Omega \rho_e(\xi)\frac{\boldsymbol{x}-\xi}{\|\boldsymbol{x}-\xi\|^3}dv(\xi) \tag{4.1}$$

4.2.2 The Definition of the Static Magnetic Field

Just as the electric field is defined in terms of the force generated by one point charge that is applied to another point charge, the magnetic field can be defined in terms of the force that one current carrying wire applies on another current carrying wire. Figure 4.2 depicts two current carrying loops of wire. The vector $\boldsymbol{x}$ locates the differential element $\boldsymbol{dl}(\boldsymbol{x})$ in the wire carrying the current $i(\boldsymbol{x})$, and the vector ξ locates the differential element $\boldsymbol{dl}(\xi)$ in the loop carrying the current $i(\xi)$. The force on the loop carrying the current $\boldsymbol{i}(\boldsymbol{x})$ is given by the expression

$$\boldsymbol{F} := \frac{\mu_0}{4\pi}\oint i(\boldsymbol{x})\boldsymbol{dl}(\boldsymbol{x}) \times \oint i(\xi)\boldsymbol{dl}(\xi) \times \frac{\boldsymbol{x}-\xi}{\|\boldsymbol{x}-\xi\|^3}.$$

The constant $\mu_0 = 4\pi \times 10^{-7}\mathrm{N/A^2}$ is the permeability of free space. The magnetic field $\boldsymbol{B}$ at the point $\boldsymbol{x}$ due to the wire carrying the current $i(\xi)$ is then defined as

$$\boldsymbol{B}(\boldsymbol{x}) := \frac{\mu_0}{4\pi}\oint i(\xi)\boldsymbol{dl}(\xi) \times \frac{\boldsymbol{x}-\xi}{\|\boldsymbol{x}-\xi\|^3}, \tag{4.2}$$

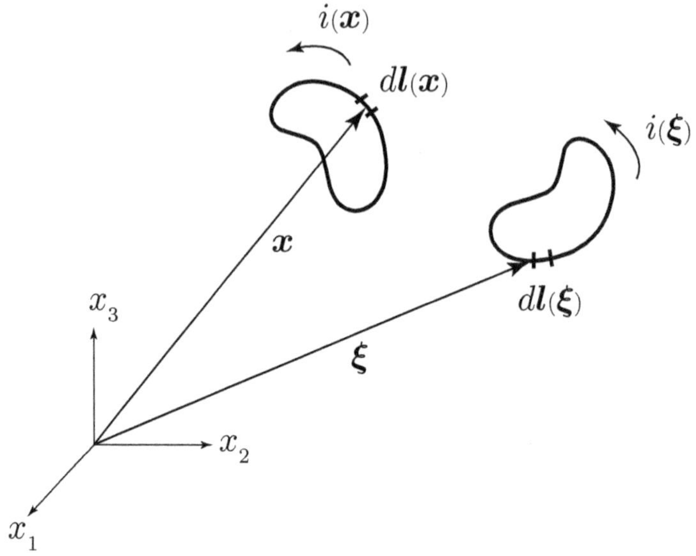

Figure 4.2 Wire loops carrying the currents $i(\boldsymbol{x})$ and $i(\xi)$, vectors $\boldsymbol{x}$ and ξ, and differential elements $d\boldsymbol{l}(\boldsymbol{x})$ and $d\boldsymbol{l}(\xi)$.

and the force acting on the loop carrying the current $i(\boldsymbol{x})$ is subsequently written as

$$\boldsymbol{F} = \oint i(\boldsymbol{x}) d\boldsymbol{l}(\boldsymbol{x}) \times \boldsymbol{B}(\boldsymbol{x}). \tag{4.3}$$

Equations 4.2 and 4.3 have been written based on the force induced by one current carrying wire on another current carrying wire. When we study electrical continua, the current is not necessarily restricted to some ideal, one dimensional domain. Rather, the current density $\boldsymbol{j}$ describes the flow of current in an arbitrary three dimensional domain Ω. Based on Equations 4.2 and 4.3, the general definition of the magnetic field is given as

$$\boldsymbol{B}(\boldsymbol{x}) := \frac{\mu_0}{4\pi} \int \boldsymbol{j}(\xi) \times \frac{\boldsymbol{x} - \xi}{\|\boldsymbol{x} - \xi\|^3} dv(\xi) \tag{4.4}$$

The magnetic field $\boldsymbol{B}$ is given in Tesla (T) in SI units. The Tesla is defined as $1\text{T} = 1\text{N}/(\text{m} \cdot \text{A})$.

4.3 Maxwell's Equations

Electrostatics is the study of the electric field $\boldsymbol{E}(\boldsymbol{x})$ given in Equation 4.1, and magnetostatics is the study of the magnetic field $\boldsymbol{B}(\boldsymbol{x})$ given in Equation 4.4.

The dynamics of the time-varying fields $\boldsymbol{E}(\boldsymbol{x}, t)$ and $\boldsymbol{B}(\boldsymbol{x}, t)$ are coupled and satisfy Maxwell's equations

$$\begin{aligned} \nabla \cdot \boldsymbol{E} &= \frac{\rho_e}{\varepsilon_0} & \nabla \cdot \boldsymbol{B} &= 0 \\ \nabla \times \boldsymbol{E} &= -\frac{\partial \boldsymbol{B}}{\partial t} & \nabla \times \boldsymbol{B} &= \mu_0 \boldsymbol{j} + \frac{1}{c^2}\frac{\partial \boldsymbol{E}}{\partial t} \end{aligned} \tag{4.5}$$

in SI units. In these equations, ε_0 is the permittivity of free space, μ_0 is the permeability of the free space, $\boldsymbol{j}$ is the current density, and ρ_e is the charge density. The constants $\varepsilon_0 = 8.854 \times 10^{-12}$F/m and $\mu_0 = 4\pi \times 10^{-7}$N/A^2. Sometimes Maxwell's equations in SI units are expressed in the slightly modified form where we replace $\varepsilon_0\mu_0 = 1/c^2$ in the equations above. It should be noted that the form of Maxwell's equations in Equation 4.5 is expressed in terms of the total charge density ρ_e and total current density $\boldsymbol{j}$. This form is sometimes known as the microscopic form of Maxwell's equations. Section 4.3.3 introduces the macroscopic form of Maxwell's equations that is tailored to electronic materials.

4.3.1 Polarization and Electric Displacement

Piezoelectric materials are examples of dielectric or polar materials. This material class is characterized by the fact that when an external electric field is applied, the resulting rearrangement of the internal charge density is constrained or limited to some lattice. Another way of stating this condition is that if we decompose the total charge density ρ_e in a domain into bound charge density $\rho_{e,b}$ and free charge density $\rho_{e,f}$ as described in Section 4.1, the free charge density $\rho_{e,f}$ is approximately zero, $\rho_{e,f} \approx 0$. Some authors use this property to define an ideal dielectric as one in which $\rho_{e,f} = 0$. Earlier we defined an insulator as a material for which $\rho_{e,f} = 0$. This means that every dielectric is an insulator but the converse is not true. In studying dielectrics we can often view the limits on the rearrangement of internal charge density to an external electric field as arising from the fixed loci of positive and negative charge in a crystal lattice.

The polarization vector $\boldsymbol{P}$ is introduced to quantify the ability of the bound charge in a dielectric to rearrange its internal charge density in response to an applied electric field. Figure 4.3 depicts a typical asymmetric crystal in which the centers of positive and negative charge do not coincide. If the magnitude of charge at each center is q, the pair of centers defines an electric dipole. The dipole moment $\boldsymbol{p}$ of the pair of charge centers having charge q is the vector

$$\boldsymbol{p} := q\boldsymbol{d},$$

where $\boldsymbol{d}$ is the vector that connects the center of negative charge to the center of positive charge. The SI units of the dipole moment are therefore C $\cdot$ m. In a material continua, and in particular for a dielectric, there will be a distribution $\boldsymbol{p}(\boldsymbol{x})$ of such electric dipoles throughout the material. We define the polarization

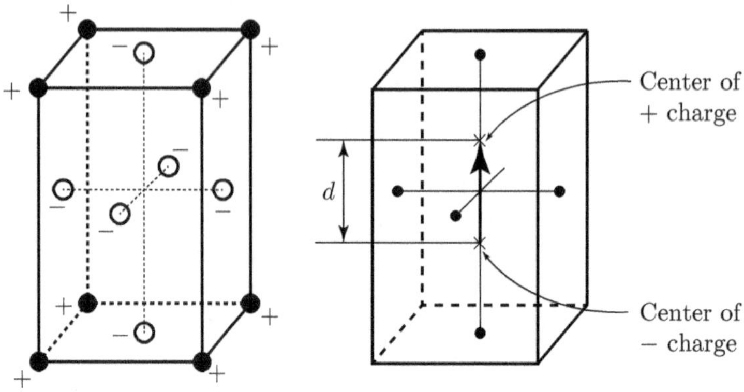

Figure 4.3 A typical crystal lattice, an asymmetric unit cell, the centers of positive and negative charge, and resulting polarization vector.

$\boldsymbol{P}(\boldsymbol{x})$ at a point $\boldsymbol{x} \in \Omega$ to be the limiting density of dipole moments per unit volume [50] in the sense that

$$\boldsymbol{P}(\boldsymbol{x}) := \frac{d\boldsymbol{p}}{dv},$$

where the limit is taken as the sample volume $\Omega(\boldsymbol{x})$ that contains the point $\boldsymbol{x}$ goes to zero. The units of the the polarization field $\boldsymbol{P}(\boldsymbol{x})$ are consequently C/m^2 in the Système Internationale. From the definition, it is apparent that the polarization does quantify one aspect of the asymmetry of the underlying crystalline lattice: if the distribution of positive and negative centers coincide everywhere in the domain Ω, then the polarization is likewise zero everywhere in Ω.

The definition of the polarization $\boldsymbol{P}$ as a density of dipole moments helps to develop an intuitive understanding of how asymmetry and polarization are intertwined. It is also important to note that the polarization vector $\boldsymbol{P}$ can be directly related to the bound charge density $\rho_{e,b}$. We always have [26, 50]

$$\rho_{e,b} = -\nabla \cdot \boldsymbol{P}.$$

From Gauss's Law, one of the constituents in Maxwell's equations 4.5, we can decompose the total charge density as $\rho_e = \rho_{e,f} - \nabla \cdot \boldsymbol{P} = \varepsilon_0 \nabla \cdot \boldsymbol{E}$. We then define the electric displacement vector

$$\boldsymbol{D} := \varepsilon_0 \boldsymbol{E} + \boldsymbol{P}, \tag{4.6}$$

from which it follows that the free charge density can be written as

$$\rho_{e,f} = \nabla \cdot \boldsymbol{D}.$$

From its definition in Equation 4.6, we see that the electric displacement $\boldsymbol{D}$ has the same units as the polarization $\boldsymbol{P}$, that is, it is given in $\mathrm{C/m^2}$ in SI.

The discussion above introduced the electric displacement $\boldsymbol{D}$ and viewed the polarization $\boldsymbol{P}$ as a means of characterizing the free charge density $\rho_{e,f}$ and bound charge density $\rho_{e,b}$, respectively. Alternatively, these fields can be interpreted as generating contributions to the total electric field $\boldsymbol{E}$: $\boldsymbol{E}_i$ is due to the redistribution of the internal charge of the material itself, and $\boldsymbol{E}_e$ is due to the external sources of charge given in

$$\boldsymbol{E} = \boldsymbol{E}_e + \boldsymbol{E}_i = \underbrace{\frac{1}{\varepsilon_0}\boldsymbol{D}}_{\text{external}} - \underbrace{\frac{1}{\varepsilon_0}\boldsymbol{P}}_{\text{internal}} .$$

The external contribution $\boldsymbol{E}_e := \frac{1}{\varepsilon_0}\boldsymbol{D}$ is the electric field that is generated by the free charge, the charge that is "wholly extraneous to the dielectric." [50] The electric field generated by the charge that accumulates on the electroded surfaces of a dielectric parallel plate capacitor is one such example. On the other hand, the self-contribution $\boldsymbol{E}_i := -\frac{1}{\varepsilon_0}\boldsymbol{P}$ represents the electric field that is generated by material charge rearrangement as the material seeks equilibrium.

The following classical example will help build intuition about the physical nature of the electric field $\boldsymbol{E}$, the free and bound charge densities $\rho_{e,f}, \rho_{e,b}$, the polarization $\boldsymbol{P}$, and the electric displacement $\boldsymbol{D}$.

Example 4.3.1 Consider the dielectric parallel plate capacitor depicted in Figure 4.4. The capacitor has thickness t, its top and bottom faces are electrodes of negligible thickness, and a potential difference of $\phi^+ - \phi^-$ is applied across the electrodes by a voltage source. We assume that the plate extends infinitely in the x_1 and x_2 directions to simplify our analysis. Suppose that the time variation of the magnetic field is negligible $\partial\boldsymbol{B}/\partial t \approx 0$, and further suppose that the dielectric is linear in the sense that $\boldsymbol{P} = \varepsilon_0\chi_e\boldsymbol{E}$ for some constant χ_e. We want to find the electric field, electric displacement, polarization and the boundary conditions on the faces of the plate.

From Maxwell's equations, the fact that $\partial\boldsymbol{B}/\partial t = 0$ implies that $\nabla \times \boldsymbol{E} = 0$. From the identity in vector calculus that $\nabla \times \nabla(\bullet) \equiv \boldsymbol{0}$, we conclude that it is possible to express the electric field in the form

$$\boldsymbol{E} = -\nabla\phi \tag{4.7}$$

for some scalar potential function ϕ. The potential function ϕ is defined rigorously in Section 4.3.4. Since we have by definition $\boldsymbol{D} := \varepsilon_0\boldsymbol{E} + \boldsymbol{P}$, it

(Continued)

Example 4.3.1 (Continued)

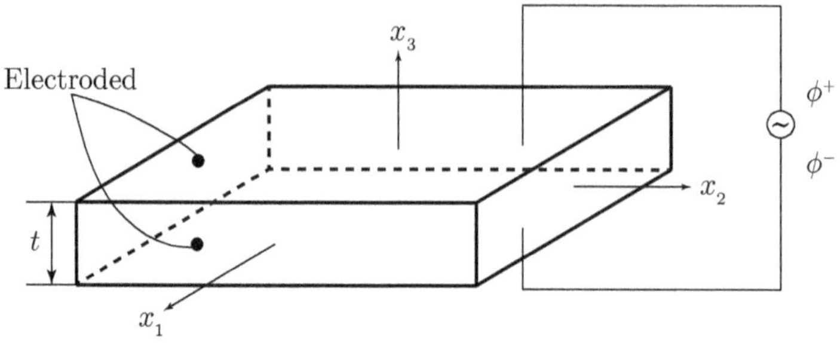

Figure 4.4 A dielectric parallel plate capacitor.

follows that

$$\boldsymbol{D} = \varepsilon \boldsymbol{E},$$

where $\varepsilon = \varepsilon_0(1 + \chi_e)$ is the relative permitivity. Since the free charge density ρ_{ef} in an ideal dielectric is zero, we conclude that

$$0 = \rho_{ef} = \nabla \cdot \boldsymbol{D} = \nabla \cdot \varepsilon \boldsymbol{E}. \tag{4.8}$$

Equations 4.7 and 4.8 can be combined to conclude that $\nabla^2 \phi = 0$ in the domain Ω. This partial differential equation is known as Laplace's equation. It is common in the study of electrostatics. A solution $\phi = \phi(x_3)$ that satisfies the boundary conditions $\phi(t/2) = \phi^+$ and $\phi(-t/2) = \phi^-$ is given by

$$\phi(x_3) = \frac{(\phi^+ - \phi^-)}{t} x_3 + \frac{1}{2}(\phi^+ + \phi^-).$$

We conclude that the electric field and electric displacement in the domain Ω satisfy

$$\boldsymbol{E} = -\nabla \phi = \begin{Bmatrix} 0 \\ 0 \\ -\dfrac{(\phi^+ - \phi^-)}{t} \end{Bmatrix} \quad \text{and} \quad \boldsymbol{D} = \varepsilon \boldsymbol{E} = \begin{Bmatrix} 0 \\ 0 \\ -\dfrac{\varepsilon(\phi^+ - \phi^-)}{t} \end{Bmatrix}.$$

We next relate the current through the voltage supply to the time-varying charge on the electrode surfaces. Consider the volume $\tilde{\Omega}$ shown in Figure 4.5 that straddles the top electrode. Using the divergence theorem of vector calculus, we know that the free charge q_{ef} contained in the volume $\tilde{\Omega}$ is given by

$$q_{ef} = \int_{\tilde{\Omega}} \rho_{ef} dv = \int_{\tilde{\Omega}} \nabla \cdot \boldsymbol{D} dv = \int_{\partial \tilde{\Omega}} \boldsymbol{D} \cdot \boldsymbol{n} da,$$

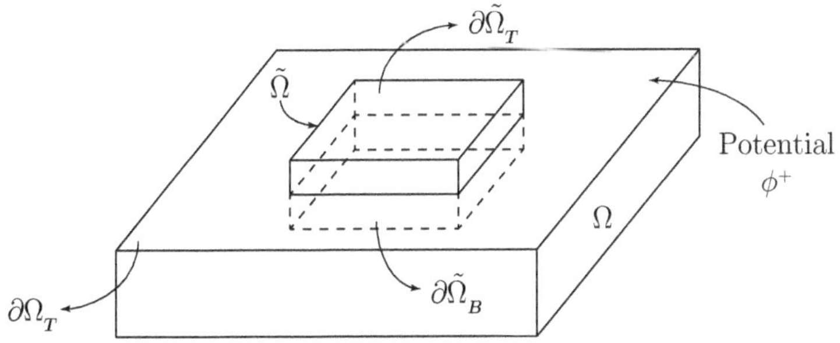

Figure 4.5 A volume $\tilde{\Omega}$ that straddles the top electrodes surface of the volume Ω.

where $\boldsymbol{n}$ is the unit outward normal at each point of the closed surface $\partial\tilde{\Omega}$ that contains $\tilde{\Omega}$. However, the total free charge in the domain $\tilde{\Omega}$ consists only of the charge on the top electrode since the free charge density ρ_{ef} is zero in the dielectric in the interior of Ω. If the surface charge density on the top electrode surface is uniform and equal to $\overline{\sigma}_T \mathrm{C/m^2}$, we have $q_{ef} = \overline{\sigma}_T A$ where A is the area of the electrode that intersects with $\tilde{\Omega}$. The calculation of the surface integral reduces to

$$\overline{\sigma}_T A = \int_{\partial\tilde{\Omega}_T} \boldsymbol{D}\cdot\boldsymbol{e}_3 da + \int_{\partial\tilde{\Omega}_B} \boldsymbol{D}\cdot(-\boldsymbol{e}_3) da$$

with e_1, e_2 and e_3 as the inertial basis. But $\boldsymbol{D}|_{\partial\tilde{\Omega}_T} = 0$ since $\boldsymbol{D} = \varepsilon\boldsymbol{E} = \boldsymbol{0}$ outside of the domain $\boldsymbol{\Omega}$. As the thickness of the volume $\tilde{\Omega}$ shrinks, we conclude that we must have

$$-D_3 = \overline{\sigma}_T \qquad \text{on } \partial\Omega_T.$$

Similar considerations, where we choose a volume $\tilde{\Omega}$ that straddles the boundary $\partial\Omega_B$, lead to the conclusion that

$$D_3 = \overline{\sigma}_B \qquad \text{on } \partial\Omega_B.$$

4.3.2 Magnetization and Magnetic Field Intensity

The polarization $\boldsymbol{P}$ and electric displacement $\boldsymbol{D}$ are used to represent the bound charge density $\rho_{e,b}$ and free charge density ρ_{ef}, respectively, in a decomposition of the charge distribution $\rho_e = \rho_{ef} + \rho_{e,b}$. A similar strategy is used to construct a decomposition of the current density $\boldsymbol{j}$. We write the current density $\boldsymbol{j}$ as

$$\boldsymbol{j} := \boldsymbol{j}_f + \boldsymbol{j}_b + \boldsymbol{j}_p \tag{4.9}$$

where $\boldsymbol{j}_f$ is the free current density, $\boldsymbol{j}_b$ is the bound current density, and $\boldsymbol{j}_p$ is the polarization current density. Since this text focuses almost exclusively on dielectrics, the free current density $\boldsymbol{j}_f$ does not appear in modeling a piezoelectric continuum *per se*. The free current density is non-zero in electroded surfaces bonded to the piezoelectrics. An expression for the bound current density $\boldsymbol{j}_b$ is constructed in terms of the magnetic polarization or simply magnetization $\boldsymbol{M}$, just as the bound charge density $\rho_{e,b}$ is expressed in terms of the polarization $\boldsymbol{P}$. We define the magnetic dipole moment m of a planar loop of current carrying wire as

$$\boldsymbol{m} := ia\boldsymbol{n}$$

where, as shown in Figure 4.6, i is the current in the wire, a is the area enclosed by the wire, and $\boldsymbol{n}$ is the unit vector normal to the plane of the wire that is generated by the right hand rule. The units of $\boldsymbol{m}$ are $A \cdot m^2$. Just as we defined the polarization $\boldsymbol{P}$, the magnetic polarization or magnetization $\boldsymbol{M}(\boldsymbol{x})$ at a point $\boldsymbol{x}$ is the density of the magnetic dipole moment per unit volume

$$\boldsymbol{M}(\boldsymbol{x}) := \frac{d\boldsymbol{m}}{dv},$$

where the limit is taken as the volume $V(\boldsymbol{x})$ containing the point $\boldsymbol{x}$ approaches zero. Consistent with the units of the magnetic dipole moment, the units of the magnetization are A/m. It can be shown (see [16], for example) that the bound current density is then written as

$$\boldsymbol{j}_b = \nabla \times \boldsymbol{M}.$$

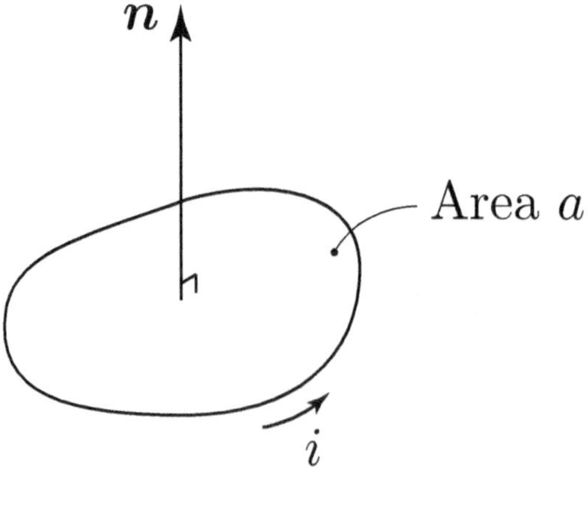

Figure 4.6 A planar loop of wire carrying current i, having area a, and normal unit vector $\boldsymbol{n}$.

The polarization current density $\boldsymbol{j}_p$ is the current density that can be attributed to time fluctuations of the electric polarization $\boldsymbol{P}$, where we have

$$\boldsymbol{j}_p = \frac{\partial \boldsymbol{P}}{\partial t}.$$

It is important to note that, in contrast to our considerations of the electric field where the bound charge density depends on $\boldsymbol{P}$, the bound current density $\boldsymbol{j}_b$ does not depend on the polarization $\boldsymbol{P}$. Rather it depends on the magnetic polarization or magnetization $\boldsymbol{M}$. The magnetic field intensity $\boldsymbol{H}$ is subsequently defined so that we have

$$\boldsymbol{B} = \mu_0(\boldsymbol{H} + \boldsymbol{M}).$$

With the introduction of the magnetization $\boldsymbol{M}$ and magnetic field intensity $\boldsymbol{H}$, Maxwell's equations can be written as follows:

$$\begin{aligned} &\nabla \cdot \boldsymbol{D} = \rho_{e,f} && \nabla \cdot \boldsymbol{B} = 0 \\ &\nabla \times \boldsymbol{E} = -\frac{\partial \boldsymbol{B}}{\partial t} && \nabla \times \boldsymbol{H} = \boldsymbol{j}_f + \frac{\partial \boldsymbol{D}}{\partial t}. \end{aligned} \tag{4.10}$$

This form of Maxwell's equations are tailored specifically to the study of electronic material continua, including dielectrics. In contrast to the case in Section 4.3, this form is expressed in terms of the free charge density $\rho_{e,f}$ and free current density $\boldsymbol{j}_f$. This form is known as the macroscopic form of Maxwell's equations.

4.3.3 Maxwell's Equations in Gaussian Units

Maxwell's equations in SI units are given in general form in Equation 4.5, and in a form that is adapted to consideration of material continua in Equation 4.10. The independent base units in the Système Internationale are the meter (m), kilogram (kg), ampere (A), and second (s). When another system of units is used, Maxwell's equations can differ in form from that given in Equations 4.5 or 4.10.

In this text, only SI or Gaussian units are considered. Gaussian units are particularly useful in developing and understanding the physical scalings that lead to the electrostatic approximation when the equations of linear piezoelectricity are derived from Maxwell's equations. Fortunately, once the equations of linear piezoelectricity are derived, their form no longer depends on the choice of which units are selected for the terms coming from Maxwell's equations. In other words, the form of the equations of linear piezoelectricity are the same whether SI or Gaussian units are used for the electrodynamic contributions.

In the Gaussian system of units, the base units for length, mass, and time are the centimeter (cm), gram (g), and second (s). There is no fourth base unit in this system of units. The general form of Maxwell's equations with respect to Gaussian units are

$$\begin{aligned}
&\nabla \cdot \boldsymbol{E} = 4\pi\rho_e && \nabla \cdot \boldsymbol{B} = 0 \\
&\nabla \times \boldsymbol{E} = -\tfrac{1}{c}\tfrac{\partial \boldsymbol{B}}{\partial t} && \nabla \times \boldsymbol{B} = \tfrac{4\pi}{c}\boldsymbol{j} + \tfrac{1}{c}\tfrac{\partial \boldsymbol{E}}{\partial t}.
\end{aligned} \tag{4.11}$$

The process by which these equations are adapted so as to be directly appropriate for continua is essentially the same that used for Maxwell's equations in SI units. By virtue of the new choice of units, however, the polarization $\boldsymbol{P}$, electric displacement $\boldsymbol{D}$, magnetization $\boldsymbol{M}$, and magnetic field intensity $\boldsymbol{H}$ are defined so that we have

$$\boldsymbol{D} = \boldsymbol{E} + 4\pi\boldsymbol{P}, \tag{4.12}$$

$$\boldsymbol{B} = \boldsymbol{H} + 4\pi\boldsymbol{M}, \tag{4.13}$$

and the charge density and current density are decomposed as

$$\rho_e = \rho_{e,f} + \rho_{e,b},$$

$$\boldsymbol{j} = \boldsymbol{j}_f + \boldsymbol{j}_b + \boldsymbol{j}_p = \boldsymbol{j}_f + c\nabla \times \boldsymbol{M} + \frac{\partial \boldsymbol{P}}{\partial t}.$$

When we substitute the charge decomposition and Equation 4.12 into Maxwell's equations 4.11, and subsequently use the identity that $-\nabla \cdot \boldsymbol{P} = \rho_{e,b}$, we find

$$\begin{aligned}
\nabla \cdot (\boldsymbol{D} - 4\pi\boldsymbol{P}) &= 4\pi(\rho_{e,f} + \rho_{e,b}), \\
\nabla \cdot \boldsymbol{D} &= 4\pi\rho_{e,f}.
\end{aligned}$$

Similarly, when the current decomposition and Equation 4.13 is substituted into Maxwell's Equations 4.11, we have

$$\begin{aligned}
\nabla \times (\boldsymbol{H} + 4\pi\boldsymbol{M}) &= \frac{4\pi}{c}\left(\boldsymbol{j}_f + c\nabla \times \boldsymbol{M} + \frac{\partial \boldsymbol{P}}{\partial t}\right) + \frac{1}{c}\frac{\partial \boldsymbol{E}}{\partial t}, \\
\nabla \times \boldsymbol{H} &= \frac{4\pi}{c}\boldsymbol{j}_f + \frac{1}{c}\frac{\partial}{\partial t}(4\pi\boldsymbol{P} + \boldsymbol{E}), \\
\nabla \times \boldsymbol{H} &= \frac{4\pi}{c}\boldsymbol{j}_f + \frac{1}{c}\frac{\partial \boldsymbol{D}}{\partial t}.
\end{aligned}$$

In summary, we can write Maxwell's equations in terms of Gaussian units for a continua in the form

$$\begin{aligned}
&\nabla \cdot \boldsymbol{D} = 4\pi\rho_{e,f} && \nabla \cdot \boldsymbol{B} = 0 \\
&\nabla \times \boldsymbol{E} = -\tfrac{1}{c}\tfrac{\partial \boldsymbol{B}}{\partial t} && \nabla \times \boldsymbol{H} = \tfrac{4\pi}{c}\boldsymbol{j}_f + \tfrac{1}{c}\tfrac{\partial \boldsymbol{D}}{\partial t}.
\end{aligned} \tag{4.14}$$

4.3.4 Scalar and Vector Potentials

Our goal in introducing Maxwell's equations is to show what physical assumptions on the electrodynamics of a continuum lead to the theory of linear piezoelectricity. This topic is studied in Chapter 5. As shown in [44], it is convenient to carry out this scaling analysis using Maxwell's equations in Gaussian units and to recast the equations in terms of scalar and vector potential functions ϕ and $\boldsymbol{\Psi}$, respectively.

We recall from vector calculus that for any vector field $\boldsymbol{A}$, we have $\nabla \cdot (\nabla \times \boldsymbol{A}) = 0$. From this general identity and Maxwell's equation $\nabla \cdot \boldsymbol{B} = 0$, we infer that it is always possible to find a vector potential function $\boldsymbol{\Psi}$ such that

$$\boldsymbol{B} = \nabla \times \boldsymbol{\Psi} \qquad \text{or} \qquad B_i = e_{ijk}\frac{\partial}{\partial x_j}\Psi_k.$$

This choice for $\boldsymbol{B}$ can be used to rewrite the specific Maxwell's equation that couples the electric and magnetic fields $\boldsymbol{E}$ and $\boldsymbol{B}$, respectively. We can write

$$\nabla \times \left(\boldsymbol{E} + \frac{1}{c}\frac{\partial \boldsymbol{\Psi}}{\partial t}\right) = 0 \qquad \text{or} \qquad e_{ijk}\frac{\partial}{\partial x_j}\left(E_k + \frac{1}{c}\frac{\partial \Psi_k}{\partial t}\right) = 0.$$

A second identity from vector calculus stipulates that for any scalar field a, we can write $\nabla \times \nabla a = 0$. The application of this identity to Maxwell's equations enables us to conclude that

$$\boldsymbol{E} + \frac{1}{c}\frac{\partial \boldsymbol{\Psi}}{\partial t} = -\nabla\phi \qquad \text{or} \qquad E_i + \frac{1}{c}\frac{\partial \Psi_i}{\partial t} = -\frac{\partial \phi}{\partial x_i}.$$

In summary, then, Maxwell's equations in terms of the vector and scalar potentials $\boldsymbol{\Psi}$ and ϕ, respectively, are then

$$\frac{\partial D_i}{\partial x_i} = 4\pi\rho_{e,f} \qquad B_i = e_{ijk}\frac{\partial \psi_k}{\partial x_j}$$

$$E_i + \frac{1}{c}\frac{\partial \psi_i}{\partial t} = -\frac{\partial \phi}{\partial x_i} \qquad e_{ijk}\frac{\partial H_k}{\partial x_j} = \frac{4\pi}{c}j_{f,i} + \frac{1}{c}\frac{\partial D_i}{\partial t},$$

where the auxiliary fields satisfy

$$D_i = E_i + 4\pi P_i,$$
$$B_i = H_i + 4\pi M_i.$$

4.4 Problems

Problems 4.4.1 Show that Maxwell's equations 4.10 for continua in SI units can be derived from the general form of Maxwell's equations 4.5 in SI units by using the decomposition of current density in Equation 4.9, the magnetization $\boldsymbol{M}$, and the definition of the magnetic field intensity $\boldsymbol{H}$.

Problems 4.4.2 Consider the lattice depicted in Figure 4.7 where a point charge of $q = 1\mu$C is located at each corner. If the dimensions of the lattice are $a = c = 1$mm and $b = 2$ mm, what is the electric field at the point $(3, 4, 5)$ mm in SI units?

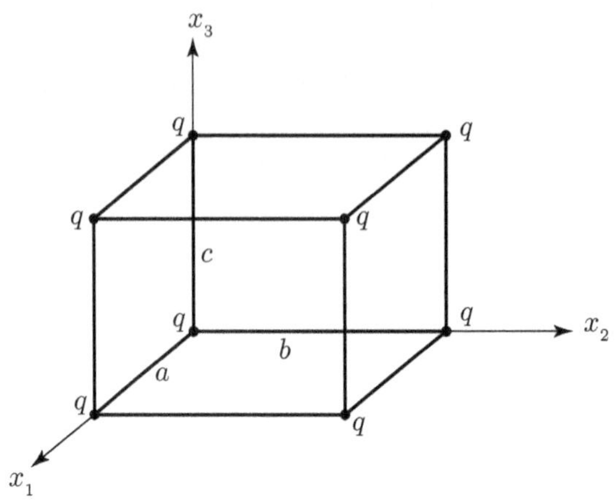

Figure 4.7 A lattice with point charges at each corner.

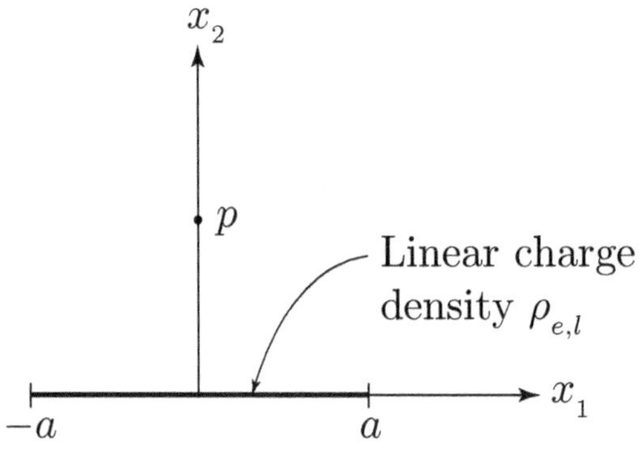

Figure 4.8 A finite wire with uniform linear charge density.

Problems 4.4.3 Consider the length of wire depicted in Figure 4.8 that carries a uniform linear charge density $\rho_{e,l} := 0.3\text{C/m}$. Find an expression for the electric field at the point p in SI units.

Problems 4.4.4 Find the electric field outside of a uniformly charged sphere of radius R that carries a total charge q. Apply the divergence theorem to Gauss's Theorem and use symmetry arguments to derive your result.

Problems 4.4.5 Find the electric field generated by an infinite plane that carries a surface charge of σ. Apply the divergence theorem to Gauss's Theorem and use symmetry arguments to derive your result.

Problems 4.4.6 If the electrostatic approximation holds, it is possible to express the electric field in the form $\boldsymbol{E} = -\nabla\phi$ where ϕ is the scalar potential. If Gauss's

Theorem is written for a domain Ω that contains no charge, it is possible to write

$$\nabla \cdot \boldsymbol{E} = \frac{\rho_e}{\varepsilon_0} = 0 \qquad \Leftrightarrow \qquad \nabla^2 \phi = 0.$$

In other words, the electric field is defined in terms of Laplace's equations for the scalar potential ϕ. Consider the empty duct having a rectangular cross section shown in Figure 4.9. The duct extends infinitely in the x_1 direction. Suppose the potential ϕ satisfies the following boundary conditions on the surface of the duct:

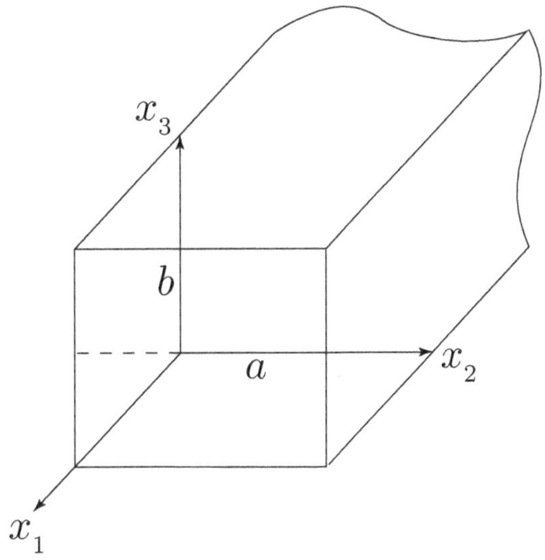

Figure 4.9 An infinite duct.

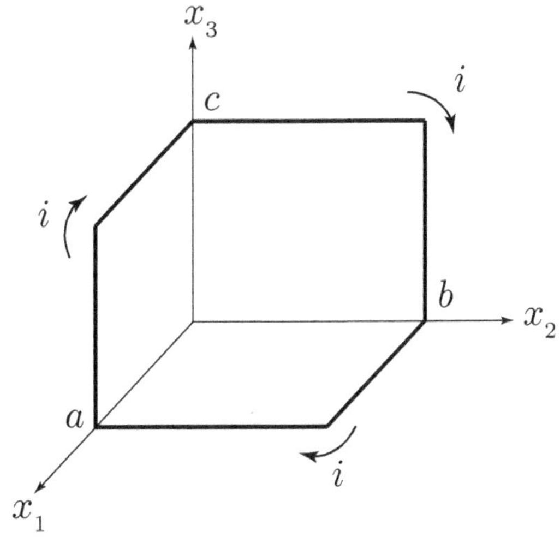

Figure 4.10 A nonplanar current carrying wire.

$$\phi|_{x_2=0} = 0$$
$$\phi|_{x_2=a} = 0$$
$$\phi|_{x_3=b} = \overline{\phi} = \text{constant}$$
$$\phi|_{x_3=-b} = \overline{\phi} = \text{constant}$$

Solve the boundary value problem for the potential function ϕ. What is the electric field in the interior of the duct?

Problems 4.4.7 Consider the non-planar current carrying loop of wire depicted in Figure 4.10. The current is 1 mA and the points a, b, c are located at 1 cm along the x_1, x_2, x_3 axes, respectively. Find the magnetic dipole moment generated by the three dimensional wire in SI units.

5

Linear Piezoelectricity

Linear piezoelectric materials transform electrical energy into mechanical energy, and vice versa, in a process that is lossless and reversible. This chapter derives the equations that govern linear piezoelectric materials in three dimensions by synthesizing results from continuum mechanics in Chapter 3 and continuum electrodynamics in Chapter 4. Simple physical experiments described in Section 5.1 illustrate how the piezoelectric effect manifests in one spatial dimension. Section 5.1 introduces the constitutive laws of linear piezoelectricity and shows how they can be interpreted as an extension of the Generalized Hooke's Law of linear elasticity. The rigorous foundations and assumptions that underly the theory of linear piezoelectricity are presented in Section 5.2. The initial-boundary value problem of linear piezoelectricity introduced in Section 5.2 is shown to be a generalization of the initial-boundary value problem of linear elasticity. Section 5.3 explores the diverse forms that the constitutive laws of linear piezoelectricity can take by using arguments based on thermodynamics.

5.1 Constitutive Laws of Linear Piezoelectricity

A macroscopic notion of how the piezoelectric effect is evident in some simple applications is illustrated in the experiments depicted in Figures 5.1 and 5.2. In Figure 5.1 a piezoelectric specimen is stretched while the top and bottom electrodes are connected by a wire. The current meter is ideal and is assumed not to affect the flow of current between the electrodes. The boundary values of the electric potential ϕ at the top and bottom electrodes are identical. If we take ϕ to be a constant as a reasonable approximation of the potential field in the domain Ω, then the electric field $E_i = -\frac{\partial \phi}{\partial x_i} = 0$. As the specimen in Figure 5.1 (left) is stretched, the electric displacement is calculated from the current measured in the meter using the boundary condition $D_3 = -\overline{\sigma}$ and the expression for the current $i = (d\overline{\sigma}/dt \cdot A)$ where $\overline{\sigma}$ is the charge density on the electrode and A is the area of the electrode.

Vibrations of Linear Piezostructures, First Edition. Andrew J. Kurdila and Pablo A. Tarazaga.

This Work is a co-publication between John Wiley & Sons Ltd and ASME Press.

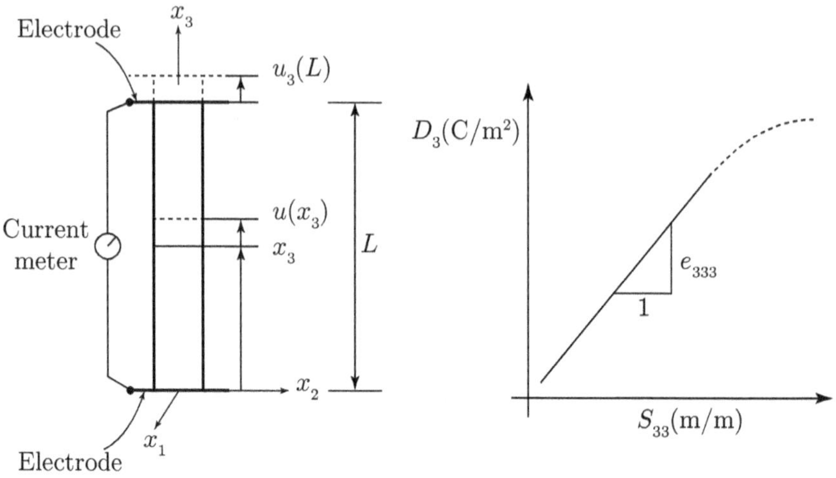

Figure 5.1 (Left) Uniaxial test to measure electric displacement, (Right) Electric displacement versus stress plot

In Figure 5.1 (right) the electric displacement D_3 is plotted versus the strain S_{33}. In a linear piezoelectric material, there is a range of response over which a linear relationship is observed between the electric displacement and the imposed strain S_{33},

$$D_3 = e_{333}S_{33}.$$

The constant of proportionality is the piezoelectric coupling constant e_{333} that has units of $\mathrm{C/m^2}$. A different experimental configuration is depicted in Figure 5.2 where the top and bottom ends of the specimen are held fixed. By symmetry we can argue that the displacement field $u_3 = 0$ in the experiment, which implies in particular that the strain $S_{33} = 0$. The voltage source generates a potential difference across the electrodes on the top and bottom of the specimen. If we assume that the potential ϕ is a linear function $\phi(x_3) = (\phi^+ - \phi^-)x_3/L + \phi^-$, the electric field is a constant $E_3 = -(\phi^+ - \phi^-)/L$. The plot in Figure 5.2 (right) depicts a typical response of a linear piezoelectric sample subject to these conditions. The stress T_{33} exhibits a linear response regime when plotted as a function of the electric field E_3 and we can write

$$T_{33} = -e_{333}E_3.$$

The constant of proportionality e_{333} is observed to be the same as that observed previously and has units of $\mathrm{C/m^2}$. The experiments summarized in Figures 5.1 and 5.2 are conducted when the electric field $E_3 \approx 0$ and the strain $S_{33} \approx 0$, respectively. Since superposition holds for linear response regimes, in a linear piezoelectric

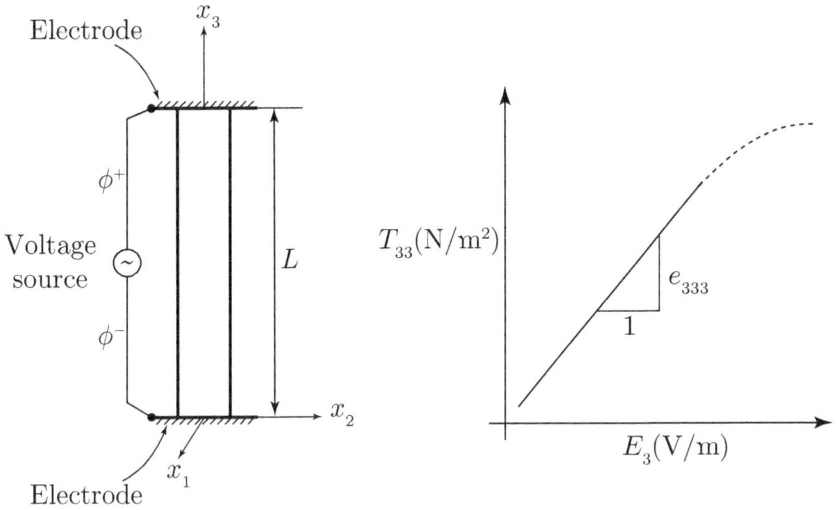

Figure 5.2 (Left) Uniaxial test to measure stress, (Right) Stress versus electric field plot

material these two independent experiments can be understood as special cases of a single, coupled one dimensional constitutive law written as

$$\begin{Bmatrix} T_{33} \\ D_3 \end{Bmatrix} = \begin{bmatrix} C^E_{3333} & -e_{333} \\ e_{333} & \epsilon^S_{33} \end{bmatrix} \begin{Bmatrix} S_{33} \\ E_3 \end{Bmatrix}.$$

In this equation, the constant C^E_{3333} is understood as the stiffness when the electric field is held constant. Similarly, ϵ^S_{33} is the permittivity when the strain is held constant.

The analysis in the previous discussion has been based on observations that are made during typical one dimensional experiments with piezoelectric materials. We noted in Section 3.4 that the generalized Hooke's law $T_{ij} = C_{ijkl}S_{kl}$ is the tensorial generalization, appropriate for understanding the relationship of stress and strain in three dimensional continua, of the one dimensional form of Hooke's law $T_{33} = C_{3333}S_{33}$ where C_{3333} is Young's modulus. In a completely analogous fashion, the general form of the components of a constitutive law for a linear piezoelectric material can be written as

$$\begin{Bmatrix} T_{ij} \\ D_m \end{Bmatrix} = \begin{bmatrix} C^E_{ijkl} & -e_{nij} \\ e_{mkl} & \epsilon^S_{mn} \end{bmatrix} \begin{Bmatrix} S_{kl} \\ E_n \end{Bmatrix}.$$

These equations relate the components T_{ij} of the second order stress tensor, S_{kl} of the second order linear strain tensor, E_n of the electric field vector, and D_m of the electric displacement vector. The coefficients C^E_{ijkl} are the components of the fourth order stiffness tensor, e_{nij} are the components of the third order piezoelectric coupling tensor, and ϵ^S_{mn} are the components of the second order permittivity tensor. In SI units the components C^E_{ijkl} are given in N/m^2, e_{nij} are given in C/m^2, and ϵ^S_{mn} are given in F/m.

As we will see in Section 5.3, there are many equivalent ways to express the same constitutive law for linearly piezoelectric materials, and the form of each law is given in terms of a set of coefficients that differ subtly from those above. For these reasons, it is common in many references to introduce superscripts to keep track among all of these various forms. For example, the superscript E in the stiffness coefficients C^E_{ijkl} is meant to emphasize that these are the coefficients that relate stress and strain when the electric field is held constant. As we will see, there are a set of *different* stiffness coefficients C^D_{ijkl} that relate stress and strain when the electric displacement is held constant. The variety of expressions for the linear constitutive laws will be discussed in more detail in Section 5.3.

5.2 The Initial-Value Boundary Problem of Linear Piezoelectricity

5.2.1 Piezoelectricity and Maxwell's Equations

While all materials conduct electricity to some degree, we noted a dielectric continuum Ω is characterized by the property that the free charge density $\rho_{e,f} \approx 0$ in Ω, so that $\rho_e = \rho_{e,b}$. Consequently, when we construct a model of piezoelectricity in a continuum Ω, we make three fundamental underlying assumptions:

(A1): The free charge density is equal to zero, $\rho_{e,f} = 0$.
(A2): The free current density is equal to zero, $\boldsymbol{j}_{e,f} = \boldsymbol{0}$.
(A3): The dielectric is nonmagnetizable, $\boldsymbol{M} = 0$.

The last assumption implies that we restrict consideration to piezoelectric materials that are nonmagnetizable. We can investigate the implications of these assumptions by considering Maxwell's equations written in terms of any set of consistent units. One immediate consequence of the assumption that $\rho_{e,f} = 0$ is that we have

$$\frac{\partial D_i}{\partial x_i} = 0. \tag{5.1}$$

This conclusion follows directly from Gauss's Theorem, the first of Maxwell's four equations. We will see that this partial differential equation for the electric displacement is one of the fundamental field equations that appear when we generalize the initial-boundary value problem of linear elasticity in Section 3.5. to obtain the initial-boundary value problem of linear piezoelectricity.

The remainder of Maxwell's equations 4.14, when they are expressed in terms of Gaussian units, reduce to the equations

$$e_{ijk}\frac{\partial H_k}{\partial x_j} = \frac{1}{c}\frac{\partial D_i}{\partial t}, \tag{5.2}$$

$$H_i = e_{ijk}\frac{\partial \psi_k}{\partial x_j}, \tag{5.3}$$

$$E_i = -\frac{\partial \phi}{\partial x_i} - \frac{1}{c}\frac{\partial \psi_i}{\partial t}, \tag{5.4}$$

where ϕ and ψ are the scalar and vector potential functions introduced in Section 4.3.4. Following the analysis in [44], we make one last assumption in addition to (A1) through (A3) above to derive the equations of piezoelectricity from Maxwell's equations. We assume

(A4): The electric field is quasistatic, $\left|\frac{\partial \phi}{\partial x_i}\right| \gg \left|\frac{1}{c}\frac{\partial \psi_i}{\partial t}\right|$.

Assumption (A4) implies in particular that the quasistatic approximation of the electric field holds, which is expressed in the form

$$E_i = -\frac{\partial \phi}{\partial x_i}. \tag{5.5}$$

Assumption (A4) implies that the electromagnetic waves essentially uncouple from the elastic waves in a piezoelectric continuum [44].

The validity of this assumption can be evaluated in terms of an iterative procedure for the solution of Maxwell's equations. The first term of the iterative approximation is constructed from the electrostatic approximation [44]. For simplicity, suppose that the continuum is a linear dielectric so that $D_i = \epsilon E_i$, where ϵ is the relative permittivity. By combining the quasistatic approximation with Gauss's Theorem in the form in Equation 5.1, the scalar potential can be found by solving Laplace's equation $\nabla^2\phi = 0$ subject to its time-varying boundary conditions. From $D_i = \epsilon E_i = -\epsilon\frac{\partial \phi}{\partial x_i}$, $\frac{\partial D_i}{\partial t}$ can subsequently be computed. An approximation of the electromagnetic wave response can then be obtained when the approximation for $\frac{\partial D}{\partial t}$ is substituted into the reduced set of Maxwell's equations 5.2 and 5.3 to obtain estimates of H_i and ψ_i. Finally, an updated approximation of the electric field $E_i = -\frac{\partial \phi}{\partial x_i} - \frac{1}{c}\frac{\partial \psi_i}{\partial t}$ can be constructed from ψ_i, and the iteration can be repeated to generate an improved estimate of the electric field. The interested reader should consult [44] for an interpretation of this iterative procedure in terms of the wave numbers of periodic solutions for the elastic waves and electromagnetic waves.

5.2.2 The Initial-Boundary Value Problem

In summary, for a piezoelectric continuum, the equations governing the electric field and electric displacement are given by Equations 5.1 and 5.5. We next give an explicit statement of the initial-boundary value problem for linear piezoelectricity that generalizes that given for linear elasticity in Section 3.5. We follow the development in [44], although the same essential proof can be found in [38]. As in Section 3.5, we assume that the boundary $\partial\Omega$ of Ω is partitioned into disjoint surfaces $\partial\Omega_T$ and $\partial\Omega_u$ that cover Ω and on which the traction and displacement boundary conditions are applied, respectively. In addition, we assume that the boundary $\partial\Omega$ is partitioned into two disjoint surfaces $\partial\Omega_\phi$ and $\partial\Omega_D$ on which electrical boundary conditions are imposed. In summary then, we have

$$\begin{aligned} \partial\Omega &= \partial\Omega_u \cup \partial\Omega_T & \qquad \partial\Omega &= \partial\Omega_\phi \cup \partial\Omega_D \\ \emptyset &= \partial\Omega_u \cap \partial\Omega_T & \qquad \emptyset &= \partial\Omega_\phi \cap \partial\Omega_D, \end{aligned}$$

as shown in Figure 5.3. In the initial-boundary value problem of linear piezoelectricity, we seek the displacement field $u_i(t,\boldsymbol{x})$, stress tensor $T_{ij}(t,\boldsymbol{x})$, linear strain tensor $S_{ij}(t,\boldsymbol{x})$, electric field $E_i(t,\boldsymbol{x})$, electric displacement $D_i(t,\boldsymbol{x})$, and scalar potential $\phi(t,\boldsymbol{x})$ that satisfy the equations

$$\begin{aligned}
\rho_m \frac{\partial^2 u_i}{\partial t^2} &= \frac{\partial T_{ji}}{\partial x_j} + f_{bi}, && \text{Equilibrium equations,} \\
S_{ij} &= \tfrac{1}{2}\left(\frac{\partial u_i}{\partial x_j} + \frac{\partial u_j}{\partial x_i}\right), && \text{Strain/displacement relationships,} \\
\begin{Bmatrix} T_{ij} \\ D_m \end{Bmatrix} &= \begin{bmatrix} C^E_{ijkl} & -e_{nij} \\ e_{mkl} & \epsilon^S_{mn} \end{bmatrix} \begin{Bmatrix} S_{kl} \\ E_n \end{Bmatrix}, && \text{Constitutive equations,} \\
\frac{\partial D_i}{\partial x_i}, &= 0 && \text{Gauss's theorem,} \\
E_i &= -\frac{\partial \phi}{\partial x_i}, && \text{Electrostatic approximation,}
\end{aligned} \tag{5.6}$$

for $(t,\boldsymbol{x}) \in \mathbb{R}^+ \times \Omega$, subject to the mechanical boundary conditions

$$\begin{aligned} T_{ij} n_j &= \overline{\tau}_i & \qquad (t,\boldsymbol{x}) &\in \mathbb{R}^+ \times \partial\Omega_T, \\ u_i &= \overline{u}_i & \qquad (t,\boldsymbol{x}) &\in \mathbb{R}^+ \times \partial\Omega_u, \end{aligned}$$

and the electrical boundary conditions

$$\begin{aligned} \phi &= \overline{\phi} & \qquad (t,\boldsymbol{x}) &\in \mathbb{R}^+ \times \partial\Omega_\phi, \\ D_j n_j &= -\overline{\sigma_e} & \qquad (t,\boldsymbol{x}) &\in \mathbb{R}^+ \times \partial\Omega_D, \end{aligned}$$

and subject to the initial conditions

$$\begin{aligned} u_i &= \overline{u}_{i,0} & \qquad \boldsymbol{x} &\in \Omega, \\ \tfrac{\partial u_i}{\partial t} &= \overline{v}_{i,0} & \qquad \boldsymbol{x} &\in \Omega, \end{aligned}$$

at $t = 0$. In these equations $\overline{\tau}_i$ is the applied surface traction acting on $\partial\Omega_T$, $\overline{u}_i$ is the displacement prescribed on $\partial\Omega_u$, $\overline{\phi}$ is the prescribed potential on $\partial\Omega_\phi$, $\overline{\sigma}$ is the

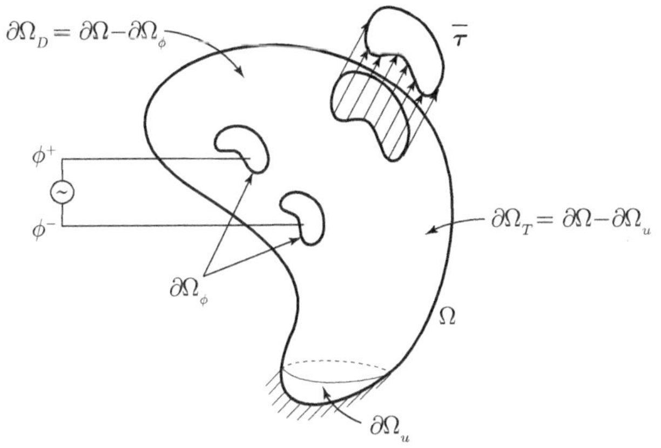

Figure 5.3 Decomposition of surface $\partial\Omega$ into complementary surfaces $\partial\Omega_u$, $\partial\Omega_T$ and $\partial\Omega_\phi$, $\partial\Omega_D$

prescribed charge distribution on $\partial\Omega_D$, $\overline{u}_0$ is the initial displacement field, and $\overline{v}_0$ is the initial velocity field. It is also important to note that the current out [47, 48] of a piezoelectric specimen is given by

$$i = -\dot{q} = -\int_{\partial\Omega_D} \dot{\overline{\sigma}} da = \int_{\partial\Omega_D} \dot{D}_j n_j da.$$

The equations for the linear piezoelectric continuum should be compared to the those derived for a linear elastic continuum in Section 3.5.

5.3 Thermodynamics of Constitutive Laws

Section 5.2.2 introduced the constitutive laws for a linearly piezoelectric continuum as a generalization based on observations made in one dimensional experiments. The result is the constitutive law that is incorporated in the initial-boundary value problem in Equations 5.6. This argument for the structure of the constitutive laws is intuitive and helps build a qualitative understanding of the physics of piezoelectric materials. As we will shortly see, however, there are a plethora of equivalent constitutive laws for linear piezoelectric materials. All of the various forms of the constitutive laws can be understood using a thermodynamic argument.

The total electromechanical energy $\mathcal{E}$ within a volume Ω of linearly piezoelectric material, which is the sum of the kinetic energy and the internal energy, is written as

$$\mathcal{E} = \frac{1}{2}\int_\Omega \rho_m \frac{\partial u_i}{\partial t}\frac{\partial u_i}{\partial t} dv + \int_\Omega \mathcal{U} dv$$

where $\mathcal{U}$ is the internal energy density. The work done by the applied surface tractions and the surface charge acting on the volume Ω has the general form

$$W = \int_{\partial\Omega} T_{ij} n_i u_j da + \int_{\partial\Omega} \phi\sigma da,$$

and it can be written in terms of the imposed boundary conditions given in the initial-boundary value problem as

$$W = \int_{\partial\Omega_u} T_{ij} n_i \overline{u}_j da + \int_{\partial\Omega_T} \overline{\tau}_i u_i da$$
$$- \int_{\partial\Omega_\phi} D_i n_i \overline{\phi} da + \int_{\partial\Omega_D} \phi\overline{\sigma} da.$$

The principle of conservation of energy states that the rate of change of the total electromechanical energy in the domain Ω is equal to the rate at which mechanical and electrical work flows across the boundary $\partial\Omega$. It follows that we have

$$\frac{d}{dt}\left\{\frac{1}{2}\int_\Omega \rho_m \frac{\partial u_i}{\partial t}\frac{\partial u_i}{\partial t} dv + \int_\Omega \mathcal{U} dv\right\} = \int_{\partial\Omega} T_{ij} n_i \frac{\partial u_j}{\partial t} da + \int_{\partial\Omega} \phi \frac{d\sigma}{dt} da.$$

By applying the divergence theorem, we can express this energy audit in terms of an integration over the volume Ω

$$\int_\Omega \left\{\rho_m \frac{\partial^2 u_i}{\partial t^2}\frac{\partial u_i}{\partial t} + \frac{d\mathcal{U}}{dt} - \frac{\partial}{\partial x_i}\left(T_{ij}\frac{\partial u_j}{\partial t}\right) + \frac{\partial}{\partial x_i}\left(\phi \frac{dD_i}{dt}\right)\right\} dv = 0,$$

and since this equality holds for any subdomain of Ω the integrand

$$\rho_m \frac{\partial^2 u_i}{\partial t^2}\frac{\partial u_i}{\partial t} + \frac{d\mathcal{U}}{dt} - \frac{\partial}{\partial x_i}\left(T_{ij}\frac{\partial u_j}{\partial t}\right) + \frac{\partial}{\partial x_i}\left(\phi \frac{dD_i}{dt}\right) = 0$$

must vanish. We impose the equilibrium conditions and Gauss's law in the expression

$$\frac{d\mathcal{U}}{dt} = \underbrace{\left(\frac{\partial T_{ji}}{\partial x_j} - \rho_m \frac{\partial^2 u_i}{\partial t^2}\right)}_{=0}\frac{\partial u_i}{\partial t} - \frac{\partial\phi}{\partial x_i}\frac{\partial D_i}{\partial t} - \phi\frac{d}{dt}\underbrace{\left(\frac{\partial D_i}{\partial x_i}\right)}_{=0} + T_{ij}\frac{\partial}{\partial t}\left(\frac{\partial u_j}{\partial x_i}\right),$$

and observe that we have the identity

$$T_{ij}\frac{\partial}{\partial t}\left(\frac{\partial u_j}{\partial x_i}\right) = \frac{1}{2}T_{ij}\frac{\partial}{\partial t}\left(\frac{\partial u_j}{\partial x_i}\right) + \frac{1}{2}T_{ji}\frac{\partial}{\partial t}\left(\frac{\partial u_j}{\partial x_i}\right) = T_{ij}\frac{\partial}{\partial t}\left(\frac{1}{2}\left(\frac{\partial u_i}{\partial x_j} + \frac{\partial u_j}{\partial x_i}\right)\right)$$
$$= T_{ij}\frac{\partial}{\partial t}(S_{ij}).$$

We conclude that the internal energy $\mathcal{U}$ satisfies the rate equation

$$\frac{d\mathcal{U}}{dt} = T_{ij}\frac{dS_{ij}}{dt} + E_i \frac{dD_i}{dt}. \tag{5.7}$$

The rate Equation 5.7 has been derived by applying a conservation law for the total electromechanical energy. It is possible, alternatively, to develop a rate equation for the internal energy from a thermodynamic viewpoint. For any linearly piezoelectric continuum, the first law of thermodynamics holds that the differential change $d\mathcal{U}$ in the internal energy $\mathcal{U}$ is equal to the sum of the differential change dQ in the head added to the system and the differential change dW in the work performed on Ω. The increment of head added can be written as $dQ = \Theta ds$ where Θ is the absolute temperature and s is the entropy, while the increment of electromechanical work is given by $dW = T_{ij}dS_{ij} + E_k dD_k$. In summary then, we have

$$\begin{aligned} d\mathcal{U} &= dQ + dW \\ &= \Theta ds + T_{ij}dS_{ij} + E_k dD_k, \end{aligned} \tag{5.8}$$

On the other hand, we can calculate the differential $d\mathcal{U}$ by assuming that the internal energy density $\mathcal{U} = \mathcal{U}(S_{ij}, D_k, s)$ is a function of the strain S_{ij}, electric displacement D_k, and entropy s. The chain rule then implies that

$$d\mathcal{U} = \left.\frac{\partial \mathcal{U}}{\partial s}\right|_{S,D} ds + \left.\frac{\partial \mathcal{U}}{\partial S_{ij}}\right|_{s,D} dS_{ij} + \left.\frac{\partial \mathcal{U}}{\partial D_k}\right|_{s,S} dD_k, \tag{5.9}$$

and a comparison of Equations 5.9 and 5.8 yields

$$\left.\frac{\partial \mathcal{U}}{\partial s}\right|_{S,D} = \Theta, \tag{5.10}$$

$$\left.\frac{\partial \mathcal{U}}{\partial S_{ij}}\right|_{s,D} = T_{ij}, \tag{5.11}$$

$$\left.\frac{\partial \mathcal{U}}{\partial D_k}\right|_{s,S} = E_k. \tag{5.12}$$

Equations 5.10, 5.11, and 5.12 can be used to derive constitutive laws for piezoelectric materials. Assume that the internal energy is a quadratic function of the strain and electric displacement

$$\mathcal{U} = \frac{1}{2} C^D_{ijkl} S_{ij} S_{kl} - h_{kij} D_k S_{ij} + \frac{1}{2} \beta^S_{ij} D_i D_j. \tag{5.13}$$

The quadratic form can alternately be written in terms of a matrix of tensors as

$$\mathcal{U} = \frac{1}{2} \left\{ S_{ij} \;\; D_m \right\} \begin{bmatrix} C^D_{ijkl} & -h_{nij} \\ -h_{mkl} & \beta^S_{mn} \end{bmatrix} \begin{Bmatrix} S_{kl} \\ D_n \end{Bmatrix}.$$

Immediately on applying Equations 5.10, 5.11, and 5.12, the linear constitutive equations have the form

$$\begin{Bmatrix} T_{ij} \\ E_m \end{Bmatrix} = \begin{bmatrix} C^D_{ijkl} & -h_{nij} \\ -h_{mkl} & \beta^S_{mn} \end{bmatrix} \begin{Bmatrix} S_{kl} \\ D_n \end{Bmatrix}. \tag{5.14}$$

The internal energy $\mathcal{U}$ is one example of what is more generally referred to as a thermodynamic potential. In addition to the internal energy $\mathcal{U}$, there are a number of other thermodynamic potentials that are useful in deriving different forms of the constitutive laws for linearly piezoelectric materials. These potentials include, for example, the Gibbs free energy and the Helmholtz free energy, among others. The interested reader can consult [43] or [26] for a discussion of all of the thermodynamic potentials. The most important of the thermodynamic potentials in this text, which is the foundation of variational approaches in Chapter 7 in deriving equations of dynamics for piezoelectric systems, is the electric enthalpy density $\mathcal{H}$. The electric enthalpy $\mathcal{H}$ is defined via the Legendre transform of the internal energy $\mathcal{U}$ and is given by

$$\mathcal{H} = \mathcal{U} - D_i E_i.$$

As we will see shortly, the Legendre transform $\mathcal{U} - D_i E_i$ of the internal energy $\mathcal{U} = \mathcal{U}(S_{ij}, D_k, s)$ induces the functional dependence $\mathcal{H} = \mathcal{H}(S_{ij}, E_k, s)$. In other words, the Legendre transform $\mathcal{U} - D_i E_i$ effectively "swaps out" the dependence of $\mathcal{U}$ on D_k so that the resulting function $\mathcal{H}$ depends on E_k. To see why this is so, we calculate the differential $d\mathcal{H}$ by the chain rule and substitute Equations 5.10, 5.11, and 5.12 to find that

$$\begin{aligned}
d\mathcal{H} &= \left.\frac{\partial \mathcal{U}}{\partial s}\right|_{D,S} ds + \left.\frac{\partial \mathcal{U}}{\partial D_i}\right|_{s,S} dD_i + \left.\frac{\partial \mathcal{U}}{\partial S_{ij}}\right|_{s,D} dS_{ij} - D_i dE_i - E_i dD_i, \\
&= \left.\frac{\partial \mathcal{U}}{\partial s}\right|_{D,S} ds + \left.\frac{\partial \mathcal{U}}{\partial S_{ij}}\right|_{s,D} dS_{ij} - D_i dE_i, \\
&= \Theta ds + T_{ij} dS_{ij} - D_i dE_i, \\
&= \left.\frac{\partial \mathcal{H}}{\partial s}\right|_{S,E} ds + \left.\frac{\partial \mathcal{H}}{\partial S_{ij}}\right|_{s,E} dS_{ij} + \left.\frac{\partial \mathcal{H}}{\partial E_i}\right|_{s,S} dE_i.
\end{aligned} \tag{5.15}$$

In analogy to Equations 5.10, 5.11, and 5.12, we conclude that

$$\left.\frac{\partial \mathcal{H}}{\partial s}\right|_{S,E} = \Theta, \tag{5.16}$$

$$\left.\frac{\partial \mathcal{H}}{\partial S_{ij}}\right|_{s,E} = T_{ij}, \tag{5.17}$$

$$\left.\frac{\partial \mathcal{H}}{\partial E_i}\right|_{s,S} = -D_i. \tag{5.18}$$

If we hypothesize a quadratic form for the electric enthalpy $\mathcal{H}$, we can use Equations 5.16, 5.17, and 5.18 to derive the form of associated constitutive laws. This process is identical to that used in the derivation of the constitutive laws in Equation 5.14. Alternatively, we can hypothesize a linear form for the constitutive laws that give T_{ij} and D_m in terms of S_{kl} and E_n and use Equations 5.16-5.17 to derive an associated quadratic form for $\mathcal{H}$. Suppose that we are given the stresses and electric displacement as a linear function of the linear strain tensor and electric field in the form

$$\begin{Bmatrix} T_{ij} \\ D_m \end{Bmatrix} = \begin{bmatrix} C^E_{ijkl} & -e_{nij} \\ e_{mkl} & \epsilon^S_{mn} \end{bmatrix} \begin{Bmatrix} S_{kl} \\ E_n \end{Bmatrix}. \tag{5.19}$$

The reader should note that this constitutive law is precisely the one introduced in Section 5.1 via intuitive arguments as an extension of the Generalized Hooke's Law. We can integrate the identity

$$\left.\frac{\partial \mathcal{H}}{\partial S_{ij}}\right|_{s,E} = C^E_{ijkl} S_{kl} - e_{mij} E_m \tag{5.20}$$

and conclude that the quadratic form for $\mathcal{H}$ has the structure

$$\mathcal{H} = \frac{1}{2} C^E_{ijkl} S_{ij} S_{kl} - e_{mij} S_{ij} E_m + f(\boldsymbol{E}) \tag{5.21}$$

for some unknown function $f = f(\boldsymbol{E})$. We next calculate the derivative of $\mathcal{H}$ with respect to the electric field and apply Equation 5.18 to find that

$$-D_i = -e_{ikl} S_{kl} + \frac{df}{dE_i} = -e_{ijk} S_{jk} - \epsilon^S_{im} E_m. \tag{5.22}$$

It is subsequently possible to integrate with respect to the electric field and find the function f. The resulting expression for the electric enthalpy density $\mathcal{H}$ is consequently

$$\mathcal{H} = \frac{1}{2} C^E_{ijkl} S_{ij} S_{kl} - e_{mij} S_{ij} E_m - \frac{1}{2} \epsilon^S_{im} E_i E_m. \tag{5.23}$$

In this section we have derived expressions for the internal energy $\mathcal{U}$ and the electric enthalpy $\mathcal{H}$ in terms of quadratic forms in Equations 5.13 and 5.23. In each case an associated set of linear constitutive relationships are obtained, Equations 5.14 and 5.19. These two sets of constitutive laws are equivalent, and in fact, it is possible to solve for one set of equations directly in terms of the other, see Problem 5.5.1. We can repeat the analysis above, relating $\mathcal{H}$ and $\mathcal{U}$ in terms of the Legendre transformation, for other choices of the thermodynamic potentials. In each case, a new thermodynamic potential is defined in terms of a Legendre transformation of some other, given thermodynamic potential. We thereby obtain a family of linear constitutive laws, each associated with a thermodynamic potential that can be expressed in terms of a quadratic form. All of the constitutive laws obtained in this fashion are in fact equivalent, and each can be expressed in terms of the others through simple algebraic calculations, such as those carried out in Problem 5.5.1. We will not have occasion to use these other forms of equivalent linear constitutive laws, or their associated thermodynamic potentials, in this text. A study of these other choices of constitutive laws can be found in [43] or [26], for example.

5.4 Symmetry of Constitutive Laws for Linear Piezoelectricity

Section 2.3 introduced how crystalline materials differ in their lattice structure and crystal class depending on their symmetry. Section 3.4 discussed how the invariance of the constitutive laws under symmetry transformations can yield important structural information about the constitutive laws. General descriptions of symmetry were used to deduce the form of the constitutive laws for monoclinic, orthotropic, and transversely isotropic linearly elastic materials. In this section, we extend this analysis to deduce how symmetry affects the structure of the constitutive laws for linearly piezoelectric materials.

As in Section 3.4, it will be most useful to perform invariance calculations using the tensor representation of the constitutive laws, but it will be more convenient to display the resulting form of the constitutive laws using the compact engineering notation. In the compact engineering notation, recall that the stress and strain tensors T_{ij}, S_{ij} are displayed in terms of vectors T_i, S_i, respectively,

$$\boldsymbol{T} := \begin{Bmatrix} T_1 \\ T_2 \\ T_3 \\ T_4 \\ T_5 \\ T_6 \end{Bmatrix} = \begin{Bmatrix} T_{11} \\ T_{22} \\ T_{33} \\ T_{23} \\ T_{13} \\ T_{12} \end{Bmatrix}, \qquad \boldsymbol{S} := \begin{Bmatrix} S_1 \\ S_2 \\ S_3 \\ S_4 \\ S_5 \\ S_6 \end{Bmatrix} = \begin{Bmatrix} S_{11} \\ S_{22} \\ S_{33} \\ 2S_{23} \\ 2S_{13} \\ 2S_{12} \end{Bmatrix}.$$

The constitutive laws for linearly piezoelectric materials can then be written in the compact matrix form

$$\begin{Bmatrix} T_i \\ D_m \end{Bmatrix} = \begin{bmatrix} C^E_{ik} & -e_{ni} \\ e_{mk} & \epsilon^S_{mn} \end{bmatrix} \begin{Bmatrix} S_k \\ E_n \end{Bmatrix},$$

where

$$C^E_{ij} = \begin{bmatrix} C^E_{11} & \cdots & C^E_{16} \\ \vdots & & \vdots \\ C^E_{16} & \cdots & C^E_{66} \end{bmatrix}, \qquad e_{ij} = \begin{bmatrix} e_{11} & \dots & e_{16} \\ e_{21} & \dots & e_{26} \\ e_{31} & \dots & e_{36} \end{bmatrix}, \qquad \epsilon^S_{ij} = \begin{bmatrix} \epsilon^S_{11} & \epsilon^S_{12} & \epsilon^S_{13} \\ \epsilon^S_{21} & \epsilon^S_{22} & \epsilon^S_{23} \\ \epsilon^S_{31} & \epsilon^S_{32} & \epsilon^S_{33} \end{bmatrix}$$

are symmetric matrices. In this book we will consider only four classes of piezoelectric materials: monoclinic C_2 or class 2 crystals, m–monoclinic crystals of class 4 or $C_s m$, trigonal crystals of class 32 or D_3, and hexagonal crystals of class 25 or C_{6v}.

5.4.1 Monoclinic C_2 Crystals

Monoclinic C_2 crystals have one of the simplest set of symmetries among the crystals. Crystals of class C_2 are invariant with respect to rotation through the angle π about an axis of symmetry [28]. The symmetry axis is referred to as the two-fold axis. If we assume that the axis of rotational symmetry is aligned with the x_1 coordinate axis, this class is invariant with respect to the rotation matrix

$$r_{ij} = \begin{bmatrix} 1 & 0 & 0 \\ 0 & -1 & 0 \\ 0 & 0 & -1 \end{bmatrix}. \tag{5.24}$$

The constitutive matrices in this case take the form [44]

$$C_{ij}^E = \begin{bmatrix} C_{11}^E & C_{12}^E & C_{13}^E & C_{14}^E & 0 & 0 \\ C_{12}^E & C_{22}^E & C_{23}^E & C_{24}^E & 0 & 0 \\ C_{13}^E & C_{23}^E & C_{33}^E & C_{34}^E & 0 & 0 \\ C_{14}^E & C_{24}^E & C_{34}^E & C_{44}^E & 0 & 0 \\ 0 & 0 & 0 & 0 & C_{55}^E & C_{56}^E \\ 0 & 0 & 0 & 0 & C_{56}^E & C_{66}^E \end{bmatrix},$$

$$e_{ij} = \begin{bmatrix} e_{11} & e_{12} & e_{13} & e_{14} & 0 & 0 \\ 0 & 0 & 0 & 0 & e_{25} & e_{26} \\ 0 & 0 & 0 & 0 & e_{35} & e_{36} \end{bmatrix},$$

$$\epsilon_{ij} = \begin{bmatrix} \epsilon_{11} & 0 & 0 \\ 0 & \epsilon_{22} & \epsilon_{23} \\ 0 & \epsilon_{23} & \epsilon_{33} \end{bmatrix}.$$

The constitutive matrices $C_{ij}^E, e_{ij}, \epsilon_{ij}$ contain 13, 8, 4 independent elements, respectively, for this crystal class.

5.4.2 Monoclinic C_s Crystals

Monoclinic crystals of class C_s also exhibit relatively straightforward symmetry properties. Class C_s crystals have a plane of symmetry, often called the mirror plane [28]. If we arrange for the mirror plane to be located in the x_2, x_3 coordinate plane so that the normal to the mirror plane is x_1, the constitutive laws are invariant with respect to the reflection transformation

$$r_{ij} = \begin{bmatrix} -1 & 0 & 0 \\ 0 & 1 & 0 \\ 0 & 0 & 1 \end{bmatrix}.$$

Note that this is an improper rotation matrix since det $(\boldsymbol{R}) = -1$. The constitutive laws for this class of crystals have the form [44]

$$C^E_{ij} = \begin{bmatrix} C^E_{11} & C^E_{12} & C^E_{13} & C^E_{14} & 0 & 0 \\ C^E_{12} & C^E_{22} & C^E_{23} & C^E_{24} & 0 & 0 \\ C^E_{13} & C^E_{23} & C^E_{33} & C^E_{34} & 0 & 0 \\ C^E_{14} & C^E_{24} & C^E_{34} & C^E_{44} & 0 & 0 \\ 0 & 0 & 0 & 0 & C^E_{55} & C^E_{56} \\ 0 & 0 & 0 & 0 & C^E_{56} & C^E_{66} \end{bmatrix},$$

$$e_{ij} = \begin{bmatrix} 0 & 0 & 0 & 0 & e_{15} & e_{16} \\ e_{21} & e_{22} & e_{23} & e_{24} & 0 & 0 \\ e_{31} & e_{32} & e_{33} & e_{34} & 0 & 0 \end{bmatrix},$$

$$\epsilon_{ij} = \begin{bmatrix} \epsilon_{11} & 0 & 0 \\ 0 & \epsilon_{22} & \epsilon_{23} \\ 0 & \epsilon_{23} & \epsilon_{33} \end{bmatrix}.$$

The constitutive matrices $C^E_{ij}, e_{ij}, \epsilon_{ij}$ contain 13, 10, 4 independent elements, respectively, for this crystal class.

5.4.3 Trigonal D_3 Crystals

The description of the symmetry of the D_3 crystal class is more complex and can be expressed in terms of five point symmetry operations, other than the identity. See [13], Table A.2 for a detailed list. Among its other symmetries, D_3 crystals have a two-fold axis of symmetry. If we arrange for the two-fold axis of symmetry to be parallel to x_1, as in our discussion of C_2 crystals, we conclude that the constitutive laws of a D_3 crystal are again invariant with respect to the rotation matrix given in Equation 5.24. It follows that the constitutive matrices for a D_3 crystal contain a subset of the independent entries of those for a monoclinic C_2 crystal. The constitutive equations for a D_3 crystal are [44]

$$C^E_{ij} = \begin{bmatrix} C^E_{11} & C^E_{12} & C^E_{13} & C^E_{14} & 0 & 0 \\ C^E_{12} & C^E_{11} & C^E_{13} & -C^E_{14} & 0 & 0 \\ C^E_{13} & C^E_{13} & C^E_{33} & 0 & 0 & 0 \\ C^E_{14} & -C^E_{14} & 0 & C^E_{44} & 0 & 0 \\ 0 & 0 & 0 & 0 & C^E_{44} & C^E_{14} \\ 0 & 0 & 0 & 0 & C^E_{14} & \frac{1}{2}(C^E_{11} - C^E_{12}) \end{bmatrix},$$

$$e_{ij} = \begin{bmatrix} e_{11} & -e_{11} & 0 & e_{14} & 0 & 0 \\ 0 & 0 & 0 & 0 & -e_{14} & -e_{11} \\ 0 & 0 & 0 & 0 & 0 & 0 \end{bmatrix},$$

$$\epsilon_{ij} = \begin{bmatrix} \epsilon_{11} & 0 & 0 \\ 0 & \epsilon_{11} & 0 \\ 0 & 0 & \epsilon_{33} \end{bmatrix}.$$

The constitutive matrices $C_{ij}^{E}, e_{ij}, \epsilon_{ij}$ contain 6, 2, 2 independent elements, respectively, for this crystal class.

5.4.4 Hexagonal C_{6v} Crystals

Among all the crystal classes that we discuss in this book, the class C_{6v} is characterized by the largest number of point symmetry operations. We see from Table A.2 in [13] that there are 11 point symmetry transformations, other than the identity, for this class. However, much of the structure of the constitutive laws can be derived by noting that this class of crystal has two orthogonal mirror planes, and the axis that is not perpendicular to these two planes is a two-fold axis of symmetry. The constitutive laws for this system can be written in the form [44]

$$C_{ij}^{E} = \begin{bmatrix} C_{11}^{E} & C_{12}^{E} & C_{13}^{E} & 0 & 0 & 0 \\ C_{12}^{E} & C_{11}^{E} & C_{13}^{E} & 0 & 0 & 0 \\ C_{13}^{E} & C_{13}^{E} & C_{33}^{E} & 0 & 0 & 0 \\ 0 & 0 & 0 & C_{44}^{E} & 0 & 0 \\ 0 & 0 & 0 & 0 & C_{44}^{E} & 0 \\ 0 & 0 & 0 & 0 & 0 & \frac{1}{2}(C_{11}^{E} - C_{12}^{E}) \end{bmatrix}, \tag{5.25}$$

$$e_{ij} = \begin{bmatrix} 0 & 0 & 0 & 0 & e_{15} & 0 \\ 0 & 0 & 0 & e_{15} & 0 & 0 \\ e_{31} & e_{31} & e_{33} & 0 & 0 & 0 \end{bmatrix},$$

$$\epsilon_{ij} = \begin{bmatrix} \epsilon_{11} & 0 & 0 \\ 0 & \epsilon_{11} & 0 \\ 0 & 0 & \epsilon_{33} \end{bmatrix}.$$

The constitutive matrices $C_{ij}^{E}, e_{ij}, \epsilon_{ij}$ contain 5, 3, 2 independent elements, respectively, for this crystal class.

5.5 Problems

Problems 5.5.1 Express the constitutive Equations 5.14 and 5.19 in matrix form using compact engineering notation. Solve for the constitutive matrices associated with 5.14 in terms to those associated with 5.19, and vice versa.

Problems 5.5.2 Show that the constitutive law for linearly piezoelectric materials of the monoclinic C_2 crystal class have the form discussed in Section 5.4.1.

Problems 5.5.3 Show that the constitutive law for linearly piezoelectric materials of the monoclinic C_s crystal class have the form discussed in Section 5.4.2.

Problems 5.5.4 Show that the constitutive law for linearly piezoelectric materials of the trigonal D_3 crystal class have the form discussed in Section 5.4.1. Refer to reference [13] Table A.2, or [28], to create a list of the symmetry transformations for the D_3 crystal class.

Problems 5.5.5 Show that the constitutive law for linearly piezoelectric materials of the hexagonal C_{6v} crystal class have the form discussed in Section 5.4.4. Refer to reference [13] Table A.2, or [28], to create a list of the symmetry transformations for the C_{6v} crystal class.

6

Newton's Method for Piezoelectric Systems

In principle, the three dimensional equations that constitute the initial-boundary value problem discussed in Section 5.2.2 can be solved to understand the coupled electrical and mechanical response of any linearly piezoelectric body. In practice, however, the number of physical systems for which an analytical solution of these equations can be derived is limited owing to their complexity. Some examples of analytical solutions to the initial-boundary value problem, mostly related to piezoelectric plate vibrations, summarized in Section 5.2.2 can be found in references [44, 47, 48]. In this chapter we show how Newton's laws of motion can be used directly to derive simple models of common piezostructural systems. This approach closely resembles the strategy employed for linearly elastic bodies that is studied in many texts on advanced strength of materials, vibrations, or structural dynamics. See [11] or [29] for a background on these methods as they are applied in vibrations or structural dynamics.

6.1 An Axial Actuator Model

In this section we derive and solve the equations of motion for a one dimensional piezoelectric actuator model. The piezoelectric composite is referred to as a "33" mode actuator because the piezoelectric coefficient e_{33} determines the degree of coupling of the voltage and strain in this actuator. Suppose the top and bottom faces of the uniaxial, linearly piezoelectric specimen of length L depicted in Figure 3.5 are electroded. A voltage source is applied between the electrodes so that the top face has potential ϕ^+ and the bottom face has potential ϕ^-. The bottom surface is constrained so that no displacement can occur, $u_3(t, 0) = 0$. The top surface is traction free so that $T_{33}(t, L) = 0$. Consider the initial-boundary value problem of linear piezoelectricity in Equation 5.6. In one dimension, in the absence of body forces, the equations of equilibrium and divergence of the electric displacement

Vibrations of Linear Piezostructures, First Edition. Andrew J. Kurdila and Pablo A. Tarazaga.

This Work is a co-publication between John Wiley & Sons Ltd and ASME Press.

can be reduced to

$$\rho_m \frac{\partial^2 u_3}{\partial t^2} = \frac{\partial}{\partial x_3}(T_{33}),$$
$$0 = \frac{\partial D_3}{\partial x_3}.$$

When we write the corresponding one dimensional constitutive law

$$T_{33} = C^E_{33} S_{33} - e_{33} E_3,$$
$$D_3 = e_{33} S_{33} + \epsilon^S_{33} E_3,$$

and introduce the electrostatic approximation $E_i = -\frac{\partial \phi}{\partial x_i}$ in terms of the scalar potential ϕ, we obtain a pair of governing partial differential equations

$$\rho_m \frac{\partial^2 u_3}{\partial t^2} = C^E_{33} \frac{\partial^2 u_3}{\partial x_3^2} + e_{33} \frac{\partial^2 \phi}{\partial x_3^2}, \tag{6.1}$$

$$0 = e_{33} \frac{\partial^2 u_3}{\partial x_3^2} - \epsilon^S_{33} \frac{\partial^2 \phi}{\partial x_3^2}. \tag{6.2}$$

These equations are subject to the boundary conditions

$$u_3(t, 0) = 0,$$
$$T_{33}(t, L) = 0,$$
$$\phi(t, 0) = 0,$$
$$\phi(t, L) = V(t),$$

for $t \in \mathbb{R}^+$ and to the initial conditions

$$u_3(0, x_3) = \overline{u}_3(x_3),$$
$$\frac{\partial u_3}{\partial t}(0, x_3) = \overline{v}_3(x_3),$$

for $x_3 \in [0, L]$ at $t = 0$. These equations govern the transient response of the electromechanical system due to an input voltage $V(t)$. Some authors refer to Equation 6.1 as the actuator equation and to Equation 6.2 as the sensor equation for this system. The next example studies the steady state response of the system to a harmonic voltage input.

Example 6.1.1 We seek a steady state solution to a harmonically driven, one dimensional, linearly piezoelectric composite by assuming that the displacement and electric fields have the form

$$u(t, x_3) = U(x_3) e^{j(\alpha t + \beta)}, \tag{6.3}$$

$$\phi(t, x_3) = \Phi(x_3) e^{j\alpha t}, \tag{6.4}$$

where α is the driving frequency of the voltage supply and β is the phase lag in the response. Approximations of this type are referred to as assumed modes methods in the vibrations and structural dynamics literature [11]. It should also be noted that the approach that follows for determining the steady state response is essentially the method of complex response that is well known in elementary vibrations [11] and summarized in Appendix S.1. When we substitute Equation 6.2 into 6.1, we obtain the wave equation

$$\rho_m \frac{\partial^2 u_3}{\partial t^2} = \left(C_{33}^E + \frac{e_{33}^2}{\epsilon_{33}^S} \right) \frac{\partial^2 u_3}{\partial x_3^2} = \bar{c} \frac{\partial^2 u_3}{\partial x_3^2}.$$

We substitute the assumed mode assumption in Equation 6.3 into the wave equation, cancel the temporal terms $e^{j(\alpha t+\beta)}$, and obtain the ordinary differential equation

$$0 = U''(x_3) + \frac{\rho_m \alpha^2}{\bar{c}} U(x_3) = U''(x_3) + \omega^2 U(x_3).$$

The solution of this equation in general is given by $U(x_3) = A\cos \omega x_3 + B \sin \omega x_3$. The second governing Equation 6.2 is integrated twice to obtain

$$e_{33} u(t, x_3) = \epsilon_{33}^S \phi(t, x_3) + C(t) x_3 + D(t).$$

By enforcing the boundary conditions $u(t, 0) = 0$ and $\phi(t, 0) = 0$, it follows that $D(t) = 0$. On substituting the assumed mode expressions in 6.3 and 6.4, we see that

$$(e_{33} U(x_3) e^{j\beta} - \epsilon_{33}^S \Phi(x_3)) e^{j\alpha t} = C(t) x_3 = \tilde{C} e^{j\alpha t} x_3 \tag{6.5}$$

where $\tilde{C}$ may be a complex constant. The stress boundary condition at $x_3 = L$ is written as

$$T_{33}(t, L) = 0 = C_{33}^E U'(L) e^{j(\alpha t+\beta)} + e_{33} \Phi'(L) e^{j\alpha t},$$

which implies that

$$0 = C_{33}^E e^{j\beta} U'(L) + e_{33} \Phi'(L). \tag{6.6}$$

Since $U, \Phi, C_{33}^E, e_{33}$ are real, it must be the case that either $\beta = 0$ or $\beta = \pi$. The boundary condition $u(t, 0) = U(0) e^{j\beta} = 0$ implies that $U(0) = 0$, which yields

$$U(x_3) = B \sin \omega x_3. \tag{6.7}$$

It also is subsequently true that the constant $\tilde{C}$ in Equation 6.5 must be real. We can combine all of these observations to conclude that

$$\Phi(x_3) = \frac{e_{33} B e^{j\beta}}{\epsilon_{33}^S} \sin \omega x_3 - \frac{\tilde{C}}{\epsilon_{33}^S} x_3 \tag{6.8}$$

(Continued)

Example 6.1.1 (Continued)

for some unknown real constants $B, \tilde{C}$. We substitute this expression and Equation 6.7 into the boundary condition 6.6 and obtain

$$0 = Be^{j\beta}\overline{c}\omega\cos\omega L - \frac{e_{33}}{\epsilon_{33}^S}\tilde{C}. \tag{6.9}$$

We also have the boundary condition

$$\phi(t, L) = \Phi(L)e^{j\alpha t} = V(t) = \overline{V}e^{j\alpha t},$$

where $\Phi(L) \equiv \overline{V}$ is the magnitude of the voltage across the electrodes of the piezoelectric specimen. Substituting Equation 6.8 into this boundary condition results in the equation

$$\overline{V} = \frac{e_{33}Be^{j\beta}}{\epsilon_{33}^S}\sin\omega L - \frac{\tilde{C}}{\epsilon_{33}^S}L. \tag{6.10}$$

The two boundary conditions in Equations 6.9 and 6.10 can be put in the matrix form

$$\begin{bmatrix} \overline{c}\omega\cos\omega L & -\frac{e_{33}}{\epsilon_{33}^S} \\ \frac{e_{33}}{\epsilon_{33}^S}\sin\omega L & -\frac{L}{\epsilon_{33}^S} \end{bmatrix} \begin{Bmatrix} Be^{j\beta} \\ \tilde{C} \end{Bmatrix} = \begin{Bmatrix} 0 \\ \overline{V} \end{Bmatrix}.$$

When this linear system of equations is solved for $B, \tilde{C}$ for a given driving frequency α, the displacement $u(t, x_3)$ and potential function $\phi(t, x_3)$ are obtained. The table below gives the physical parameters for a piezoelectric specimen that has a length of one centimeter.

Constant	Value	Units
C_{33}^E	12.6×10^{10}	N/m^2
e_{33}	23.2	C/m^2
ϵ_{33}^S	1.51×10^{-8}	$C/(V \cdot m^2)$
ρ	7500	kg/m^2
L	0.01	m
$\overline{V}$	100	V

Figure 6.1 plots the assumed mode $U(x_3)$ for the steady state response $u(t, x_3) = U(x_3)e^{j(\alpha t+\beta)}$. It is evident that the nature of the steady state solution $U(x_3)$ varies dramatically as the driving frequency is varied from 100Hz to 10MHz. As the driving frequency becomes greater, the standing wave represented by $U(x_3) = B\sin\omega x_3$ becomes more oscillatory. At 100 Hz the

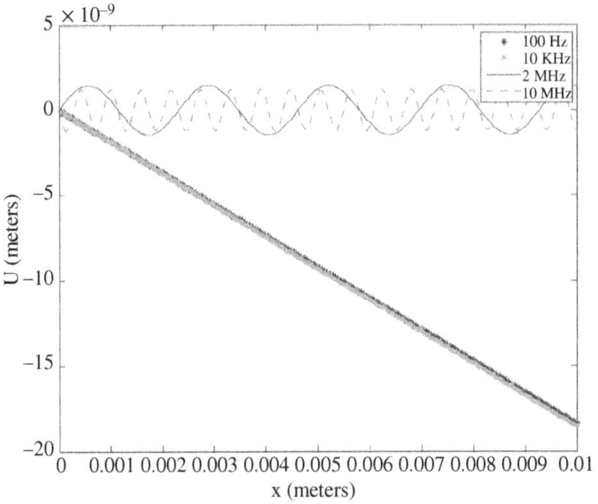

Figure 6.1 Displacement assumed mode $U(x_3)$.

frequency $\omega = \sqrt{\rho\alpha^2/\bar{c}}$ is so low that the response is much less than a single period and appears almost linear to the eye. The oscillations become ever more apparent as the driving frequency α is increased.

Figure 6.2 depicts the corresponding plot of the assumed mode $\Phi(x_3)$ of the scalar potential $\phi(t, x_3) = \Phi(x_3)e^{j\alpha t}$. From Equation 6.8 we see that

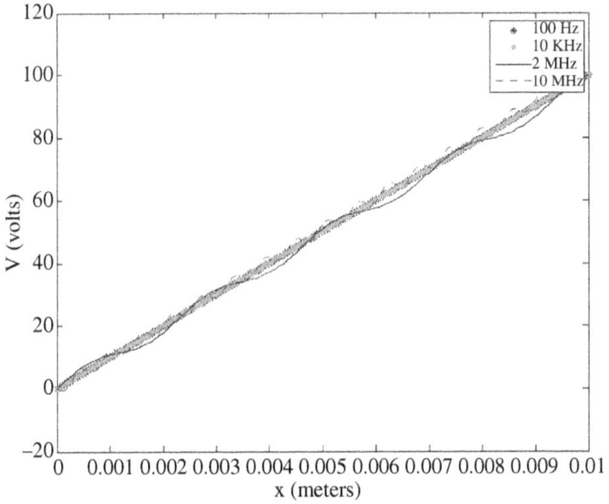

Figure 6.2 Potential assumed mode $\Phi(x_3)$.

(Continued)

Example 6.1.1 (Continued)

there are sinusoidal and linear contributions to the solution. In the figure below, we see that the linear contribution dominates the total response. At low driving frequencies α, the response is nearly linear. At high driving frequencies, even on the order of $\alpha \sim O(10\text{MHz})$, the response takes the form of small oscillations about the linear response. This observation will be critical in the sections that follow where an approximation of the potential field is used to derive several prototypical active structures.

6.2 An Axial, Linear Potential, Actuator Model

Example 6.1.1 showed that for the one dimensional geometry of a uniaxial specimen, the equations of linear piezoelectricity yield a system of equations that couple the displacement and electrical field along the axial direction. For harmonic input voltage time histories, the steady state solution of this coupled system is shown to yield an assumed mode $\Phi(x_3)$ for the potential $\phi(t, x_3) = \Phi(x_3)e^{j\alpha t}$ that is approximately a linear function in x_3 over wide ranges of the voltage $V(t) = \overline{V}e^{j\alpha t}$. In this section we introduce a common approximation based on this observation that is often used to obtain a simpler governing equation for the deformation in prototypical piezoelectric composite structures.

As shown in Figure 6.3, the uniaxial element is oriented along the x_3 direction. As is typical for a ferroelectric, it is assumed that the material has been poled so that the direction of polarization is likewise along x_3. The potential $\phi = \phi(t, x_3)$ has values $\phi^+(t)$ and $\phi^-(t)$, respectively, on the electroded top and bottom surfaces of the specimen. The thickness of the electrode is assumed to be negligible and does not contribute to the mechanical properties of the specimen. In contrast to Example 6.1.1, in this example we approximate the potential field ϕ as a linear function of the variable x_3 alone,

$$\phi(t, x_3) := (\phi^+(t) - \phi^-(t))\frac{x_3}{L} + \phi^-(t) = \frac{V(t)}{L}x_3 + \phi^-(t).$$

Since the electric field is given by $E_i = -\frac{\partial \phi}{\partial x_i}$, we consequently have

$$E_3(x_3) = -\frac{V}{L}\chi_{[0,L]}(x_3)$$

where the characteristic function $\chi_{[0,L]}(x_3) = 1$ whenever $x_3 \in [0, L]$ and is equal to zero otherwise. Because of the simple geometry of the problem at hand, we assume that the constitutive law takes the form

$$\begin{Bmatrix} T_{33} \\ D_3 \end{Bmatrix} = \begin{bmatrix} C_{33}^E & -e_{333} \\ e_{333} & \epsilon_{33}^S \end{bmatrix} \begin{Bmatrix} S_{33} \\ E_3 \end{Bmatrix}.$$

When this constitutive model holds, we say that the specimen undergoes '33 mode' deformation. The electrical field variables and the mechanical field variables both are assumed to undergo variations along the 3– axis. We extract a slice of thickness dx_3 that is located at the x_3 coordinate as depicted in Figure 6.3. Newton's equations of motion along the x_3 direction yield

$$\begin{aligned} \rho_m A(x_3)dx_3 \frac{\partial^2 u_3}{\partial t^2} &= T_{33}(x_3 + dx_3)A(x_3 + dx_3) - T_{33}(x_3)A(x_3) \\ &= \left(T_{33}(x_3) + \frac{\partial T_{33}}{\partial x_3} dx_3 \right) \left(A(x_3) + \frac{\partial A}{\partial x_3} dx_3 \right) - T_{33}(x_3)A(x_3) + O(dx_3^2). \end{aligned}$$

In this equation $T_{33}(x_3 + dx_3)$ and $A(x_3 + dx_3)$ are the normal stress and area of the top face of the slice, respectively. The quantities $T_{33}(x_3)$ and $A(x_3)$ are the corresponding terms for the bottom of the slice. If we divide both sides of this equation by dx_3 and take the limit as $dx_3 \to 0$, the governing equation can be written as

$$\rho_m A \frac{\partial^2 u_3}{\partial t^2} = \frac{\partial}{\partial x_3}(AT_{33}).$$

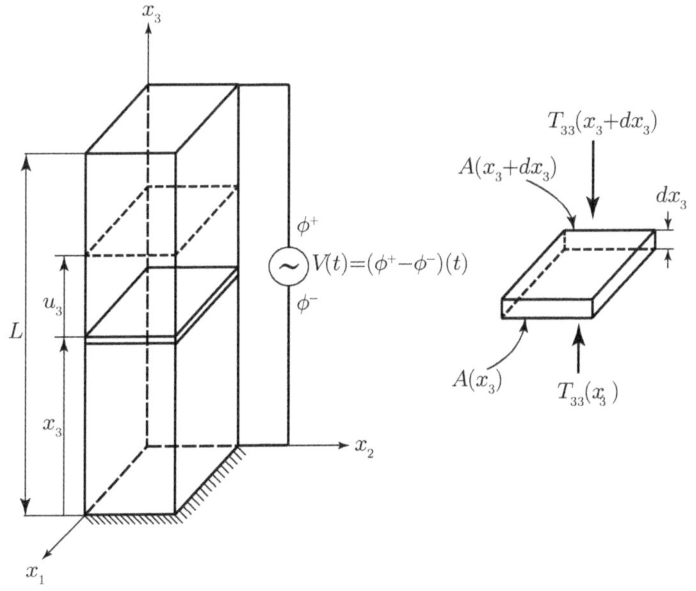

Figure 6.3 Piezoelectric uniaxial rod with the top and bottom surfaces electroded.

The final form of the governing equation is obtained when we substitute the first row of the constitutive law into Equation 6.2 and subsequently introduce the strain-displacement relationship $S_{33} = \frac{\partial u_3}{\partial x_3}$. We thereby obtain an evolutionary equation in which we seek $u_3 = u_3(t,x)$ that satisfies

$$\rho_m A \frac{\partial^2 u_3}{\partial t^2} = \frac{\partial}{\partial x_3}\left(AC^E_{33}\frac{\partial u_3}{\partial x_3}\right) + \frac{\partial}{\partial x_3}\left(\frac{Ae_{33}}{L}\chi_{[0,L]}\right)V \tag{6.11}$$

for $(t,x_3) \in \mathbb{R} \times [0,L]$, subject to the boundary conditions

$$u(t,0) = 0$$
$$T_{33}(t,L) = 0$$

for all $t \in \mathbb{R}^+$, and the initial conditions

$$u_3(0,x_3) = \overline{u}_3(x_3)$$
$$\frac{\partial u_3}{\partial t}(0,x_3) = \overline{v}_3(x_3)$$

for all $x_3 \in [0,L]$ at $t = 0$.

6.3 A Linear Potential, Beam Actuator

Another common architecture used to realize a solid state actuator, one that produces a displacement with the application of a voltage, is the piezoelectrically actuated composite beam shown in Figure 6.4. This composite is referred to as a bimorph actuator by some authors, and it is also described as a "31" mode actuator. The latter description is due to the fact that the e_{31} piezoelectric constant determines the coupling between the electric field in the 3-direction and the strain S_{11} in the 1-direction in the composite.

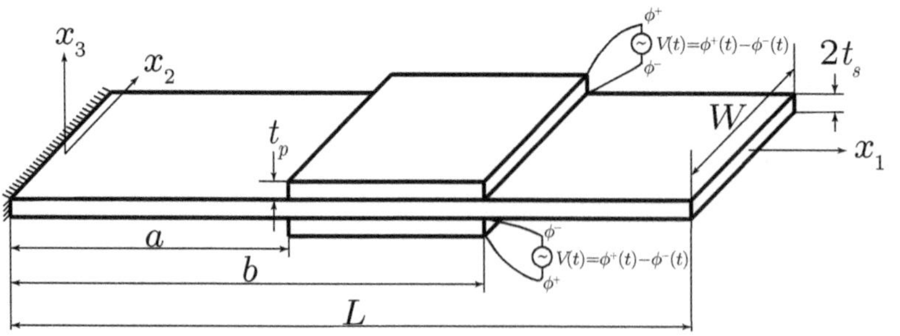

Figure 6.4 Piezoelectrically actuated composite beam.

In this configuration the top and bottom piezoelectric patches are electroded on both sides, and each electrode is assumed to be thin enough that its contribution to the mechanical properties of the composite structure are negligible. In many experimental realizations of this system, the host or substrate is assumed to be conducting. It is then clear that the top and bottom of the substrate have potential ϕ^+. If we suppose that the piezoelectric patches are in fact ferroelectric, it is assumed that the material has been poled and that the poling direction of both patches is along the x_3 direction.

In view of our analysis of the composite piezoelectric in Section 6.1, we assume that a good approximation of the electric field can be achieved by representing the potential as a piecewise linear function. We therefore approximate the potential field over the piezoelectric patch as a piecewise linear function that satisfies the potential boundary conditions.

$$\phi(t, x_1, x_3) =$$
$$\begin{cases} \dfrac{(\phi^+ - \phi^-)}{t_p} x_3 - \dfrac{t_s}{t_p}\phi^+ + \dfrac{(t_s + t_p)}{t_p}\phi^- & (x_1, x_3) \in [a, b] \times [t_s, t_s + t_p], \\ \dfrac{-(\phi^+ - \phi^-)}{t_p} x_3 - \dfrac{t_s}{t_p}\phi^+ + \dfrac{(t_s + t_p)}{t_p}\phi^- & (x_1, x_3) \in [a, b] \times [-(t_s + t_p), -t_s], \\ 0 & \text{otherwise.} \end{cases}$$

With this choice of potential field, the electric field $E_i = -\frac{\partial \phi}{\partial x_i}$ consequently is given by

$$E_3(x_1, x_3) := \begin{cases} -\dfrac{V}{t_p} & (x_1, x_3) \in [a, b] \times [t_s, t_s + t_p] \\ \dfrac{V}{t_p} & (x_1, x_3) \in [a, b] \times [-(t_s + t_p), -t_s] \end{cases} \tag{6.12}$$

The constitutive law for both piezoelectric patches is assumed to be identical since they have a common poling direction. In both cases it is assumed that the bending stress and strain, T_{11} and S_{11}, respectively, are sufficient to model the response of the composite. In addition it is assumed that the dominant electrical field variables are the electric displacement D_3 and electric field E_3. The general form of the constitutive law consequently is reduced to the simple form

$$\begin{Bmatrix} T_{11} \\ D_3 \end{Bmatrix} = \begin{bmatrix} C_{11}^E & e_{31} \\ e_{31} & \epsilon_{33}^S \end{bmatrix} \begin{Bmatrix} S_{11} \\ E_3 \end{Bmatrix}. \tag{6.13}$$

The equilibrium conditions for the beam are derived following standard techniques from linear vibrations theory or structural dynamics [11]. We extract a section of the beam as depicted in Figure 6.5 and apply Newton's Laws of motion

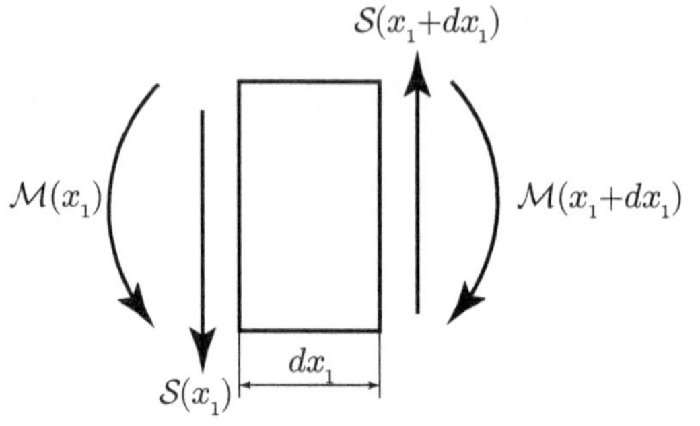

Figure 6.5 Shear and bending moment acting on a typical beam section.

to the resulting free body diagram. The bending moment $\mathcal{M}$ and shear force S shown in the figure are defined via

$$\mathcal{M}(x_1) := \iint x_3 T_{11} dx_2 dx_3, \tag{6.14}$$

$$S(x_1) := \int T_{13} dx_2 dx_3. \tag{6.15}$$

With these definitions of the stress resultants, the shear force S on the right face of the section acts in the positive x_3 direction, while the bending moment $\mathcal{M}$ acting on the right face acts in the clockwise direction. When we sum forces and moments, we obtain two equations

$$\begin{aligned}\rho_m A dx_1 \frac{\partial^2 u_3}{\partial t^2} &= S(x_1 + dx_1) - S(x_1) \\ &= S(x_1) + \frac{\partial S}{\partial x_1} dx_1 - S(x_1) + O(dx_1)^2,\end{aligned}$$

and

$$\begin{aligned}0 &= -\mathcal{M}(x_1 + dx_1) + \mathcal{M}(x_1) + S(x_1 + dx_1)dx_1 \\ &= -\mathcal{M}(x_1) - \frac{\partial \mathcal{M}}{\partial x_1} dx_1 + \mathcal{M}(x_1) + S(x_1)dx_1 + O(dx_1)^2.\end{aligned}$$

We subsequently divide by dx_1 and take the limit as $dx_1 \to 0$. The equilibrium conditions then take the form familiar from classical linear vibrations or structural dynamics where we have

$$\rho_m A \frac{\partial^2 u_3}{\partial t^2} = \frac{\partial S}{\partial x_1}, \tag{6.16}$$

$$S = \frac{\partial \mathcal{M}}{\partial x_1}. \tag{6.17}$$

Recall from Chapter 3 in Equation 3.3 that the bending strain in a Bernoulli–Euler beam takes the form

$$S_{11} \approx -x_3 \frac{\partial^2 u_3}{\partial x_1^2}. \tag{6.18}$$

When the strain-displacement relationship in Equation 6.18 and the stress from the simplified constitutive law in Equation 6.13 are substituted into the definition of the moment in Equation 6.14, we obtain

$$\begin{aligned}\mathcal{M} &= \iint x_3(C_{11}^E S_{11} - e_{31}E_3)dx_2dx_3 \\ &= -\iint C_{11}^E x_3^2 da \frac{\partial^2 u_3}{\partial x_1^2} - \iint x_3 e_{31} E_3 dx_2 dx_3 \end{aligned} \tag{6.19}$$

where the term

$$(C_{11}^E I)(x_1) := \iint C_{11}^E x_3^2 da$$

is the classical definition of the bending stiffness of the beam and I is the area moment of inertia. We exploit the symmetry of the cross section and the expression for the electric field in Equation 6.12 to define a simple expression for the rightmost term in Equation 6.19.

$$\begin{aligned}\iint x_3 e_{31} E_3 dx_2 dx_3 &= \left(-\int\int_{\text{Top}} x_3 dx_2 dx_3 + \int\int_{\text{Bottom}} x_3 dx_2 dx_3\right)\frac{e_{31}\chi_{[a,b]}}{t_p}V \\ &= (-K_T + K_B)\frac{e_{31}\chi_{[a,b]}}{t_p}V \\ &= -\frac{Ke_{31}\chi_{[a,b]}}{t_p}V\end{aligned}$$

The constant $K := K_T - K_B$ is known as the area moment of the entire cross section, while K_T and $-K_B$ are the area moments of the top and bottom portions of the cross section. These constants are easily computed in closed form for simple cross sectional geometries. The final form of the evolution equation that governs the displacement $u_3 = u_3(t, x_1)$ is obtained by substituting the definition of the area moments K_T, K_B, the bending stiffness $C_{11}^E I$, and bending moment $\mathcal{M}$ into the equilibrium Equations 6.16 and 6.17. In summary, we seek a $u_3(t, x_1)$ that satisfies the evolution equation

$$\rho_m A \frac{\partial^2 u_3}{\partial t^2} = -\frac{\partial^2}{\partial x_1^2}\left(C_{11}^E I \frac{\partial^2 u_3}{\partial x_1^2}\right) + \frac{\partial^2}{\partial x_1^2}\left(\frac{Ke_{31}\chi_{[a,b]}}{t_p}\right)V \tag{6.20}$$

for all $(t, x_1) \in \mathbb{R}^+ \times [0, L]$, subject to the boundary conditions at $x_1 = 0$

$$u_3(t,0) = \overline{u}_{3,0}(t) \quad \text{or} \quad \mathcal{S}(t,0) = \overline{\mathcal{S}}_0(t),$$

$$\frac{\partial u_3}{\partial x_1}(t,0) = \overline{\theta}_0(t) \quad \text{or} \quad \mathcal{M}(t,0) = \overline{\mathcal{M}}_0(t),$$

and the boundary conditions at $x_1 = L$

$$u_3(t, L) = \overline{u}_{3,L}(t) \qquad \text{or} \qquad \mathcal{S}(t, L) = \overline{\mathcal{S}}_L(t),$$

$$\frac{\partial u_3}{\partial x_1}(t, L) = \overline{\theta}_L(t) \qquad \text{or} \qquad \mathcal{M}(t, L) = \overline{\mathcal{M}}_L(t),$$

and subject to the initial conditions

$$u_3(0, x_1) = \overline{u}_3(x_1),$$

$$\frac{\partial u_3}{\partial t}(0, x_1) = \overline{v}_3(x_1)$$

for $x_1 \in [0, L]$ at $t = 0$.

6.4 Composite Plate Bending

The piezoelectrically actuated beam described in Section 6.3 is one of the most popular active structural elements and is employed in numerous applications. In some applications, we require bending in two perpendicular directions and the simple structure in Section 6.3 does not suffice. This section discusses the mechanics of the piezoelectrically actuated composite plate depicted in Figure 6.6. We refer to the central layer as the host or substrate layer. Depending on the specific model, the host layer may be linearly elastic or linearly piezoelectric. Two thin piezoelectric patches are bonded to opposite surfaces of the host layer. It is assumed that the composite structure satisfies the hypotheses of Kirchoff plate theory. Specifically, it is assumed that (1) the thickness of the plate is much less than the local radius of curvature anywhere on the deformed plate, (2) the magnitude of the normal stress T_{33} in the transverse direction is much less than the in-plane normal stresses T_{11} and T_{22}, (3) fibers normal to the neutral surface remain normal during deformation, and (4) the local radius of curvature in the x_1 and x_2 directions

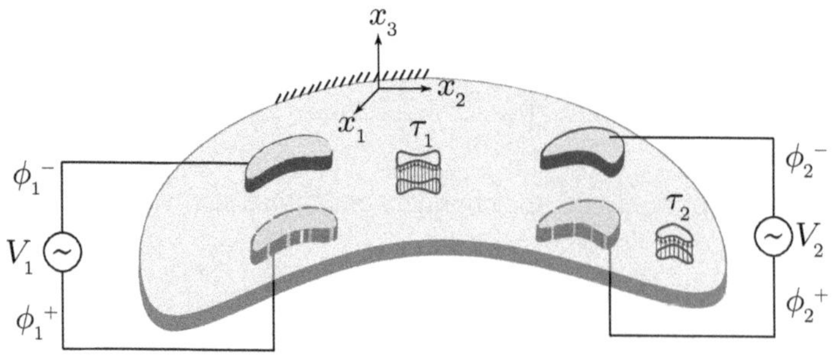

Figure 6.6 Piezoelectrically actuated composite plate driven by two voltage sources.

can be approximated by linear second order differential operator. A discussion of these assumptions can be found in many classical texts on mechanics, see [4] for example.

The derivation of the equations of motion for the plate begins by defining the stress resultants that act along a typical edge of the plate. The shear forces per unit length are given by the expressions

$$S_1 = \int_{-h/2}^{h/2} T_{13} dx_3 \qquad S_2 = \int_{-h/2}^{h/2} T_{23} dx_3, \tag{6.21}$$

while the bending moments per unit length are defined as

$$\begin{aligned} \mathcal{M}_1 &= \int_{-h/2}^{h/2} x_3 T_{11} dx_3 & \mathcal{M}_2 &= \int_{-h/2}^{h/2} x_3 T_{22} dx_3 \\ \mathcal{M}_{12} &= \int_{-h/2}^{h/2} x_3 T_{12} dx_3 & \mathcal{M}_{21} &= \int_{-h/2}^{h/2} x_3 T_{21} dx_3. \end{aligned} \tag{6.22}$$

These quantities should be compared to the bending moment $\mathcal{M}$ and shear force S defined for the Bernoulli–Euler beam in Equations 6.14 and 6.15, respectively. We emphasize that, in contrast to the definitions for bending moment $\mathcal{M}$ and shear force S for the Bernoulli–Euler beam, the stress resultants $S_1, S_2, \mathcal{M}_1, \mathcal{M}_2, \mathcal{M}_{12}, \mathcal{M}_{21}$ for the Kirchoff plate are defined as shear or moment per unit length.

When we carry out a force summation in the x_3 direction, we find that

$$\begin{aligned} \rho dx_1 dx_2 h \frac{\partial^2 u_3}{\partial t^2} &= \left(S_1 + \frac{\partial S_1}{\partial x_2} dx_1 \right) dx_2 - S_1 dx_2 + \left(S_2 + \frac{\partial S_2}{\partial x_2} \right) dx_1 - S_2 dx_1 \\ &\quad + dx_1 dx_2 \mathcal{O}(dx_i). \end{aligned}$$

We divide both sides of the resulting equation by $dx_1 dx_2$ and take the limit as $dx_1 \rightarrow 0, dx_2 \rightarrow 0$ to obtain the transverse displacement equation.

$$\rho h \frac{\partial^2 u_3}{\partial t^2} = \frac{\partial S_1}{\partial x_2} + \frac{\partial S_2}{\partial x_2}. \tag{6.23}$$

We next sum moments about the x_1 axis and find that

$$\begin{aligned} 0 = \mathcal{M}_2 dx_1 &- \left(\mathcal{M}_2 + \frac{\partial \mathcal{M}_2}{\partial x_2} dx_2 \right) dx_1 + \mathcal{M}_{12} dx_2 \\ &- \left(\mathcal{M}_{12} + \frac{\partial \mathcal{M}_{12}}{\partial x_1} dx_1 \right) dx_2 - S_1 dx_2 \frac{dx_2}{2} \\ &+ \left(S_1 + \frac{\partial S_1}{\partial x_1} dx_1 \right) dx_2 \frac{dx_2}{2} + \left(S_2 + \frac{\partial S_2}{\partial x_2} dx_2 \right) dx_1 dx_2 \\ &+ dx_1 dx_2 \mathcal{O}(dx_i), \end{aligned}$$

which implies, when we divide by $dx_1 dx_2$ and take the limit as $dx_i \to 0$, that

$$S_2 = \frac{\partial \mathcal{M}_{21}}{\partial x_1} + \frac{\partial \mathcal{M}_2}{\partial x_2}. \tag{6.24}$$

A similar analysis of the moments about the x_2 axis yields

$$S_1 = \frac{\partial \mathcal{M}_1}{\partial x_1} + \frac{\partial \mathcal{M}_{21}}{\partial x_2}. \tag{6.25}$$

The final form of the equilibrium equation for the lateral deformation of the plate results when we substitute Equations 6.24, 6.25 into Equation 6.23:

$$\rho h \frac{\partial^2 u_3}{\partial t^2} = \frac{\partial^2 \mathcal{M}_1}{\partial x_1^2} + \frac{\partial^2 \mathcal{M}_2}{\partial x_2^2} + \frac{\partial^2 \mathcal{M}_{12}}{\partial x_2 x_1} + \frac{\partial^2 \mathcal{M}_{21}}{\partial x_1 x_2}. \tag{6.26}$$

It should be noted that Equation 6.26 is a generic model of a Kirchoff plate; it holds if the plate is linearly elastic or linearly piezoelectric. In this section we assume that the piezoelectric material is of the hexagonal crystal class. The full constitutive law can therefore be written in compact engineering notation in the form

$$\left\{ \begin{Bmatrix} T_1 \\ T_2 \\ T_3 \\ T_4 \\ T_5 \\ T_6 \end{Bmatrix} \atop \begin{Bmatrix} D_1 \\ D_2 \\ D_3 \end{Bmatrix} \right\} = \begin{bmatrix} \begin{bmatrix} C_{11}^E & C_{12}^E & C_{13}^E & 0 & 0 & 0 \\ C_{12}^E & C_{22}^E & C_{23}^E & 0 & 0 & 0 \\ C_{13}^E & C_{23}^e & C_{33}^E & 0 & 0 & 0 \\ 0 & 0 & 0 & C_{44}^E & 0 & 0 \\ 0 & 0 & 0 & 0 & C_{55}^E & 0 \\ 0 & 0 & 0 & 0 & 0 & C_{66}^E \end{bmatrix} & \begin{bmatrix} 0 & 0 & -e_{31} \\ 0 & 0 & -e_{32} \\ 0 & 0 & -e_{33} \\ 0 & -e_{24} & 0 \\ -e_{15} & 0 & 0 \\ 0 & 0 & 0 \end{bmatrix} \\ \begin{bmatrix} 0 & 0 & 0 & 0 & e_{15} & 0 \\ 0 & 0 & 0 & e_{24} & 0 & 0 \\ e_{31} & e_{32} & e_{33} & 0 & 0 & 0 \end{bmatrix} & \begin{bmatrix} \epsilon_{11} & 0 & 0 \\ 0 & \epsilon_{22} & 0 \\ 0 & 0 & \epsilon_{33} \end{bmatrix} \end{bmatrix} \left\{ \begin{Bmatrix} S_1 \\ S_2 \\ S_3 \\ S_4 \\ S_5 \\ S_6 \end{Bmatrix} \atop \begin{Bmatrix} E_1 \\ E_2 \\ E_3 \end{Bmatrix} \right\}. \tag{6.27}$$

We know that $T_{33} = 0$ on the surfaces of the plate. Since the plate is thin, we consequently approximate $T_{33} \approx 0$ throughout the thickness of the plate. The third row in Equation 6.27 exploits this fact and defines a constraint

$$S_3 = \frac{1}{C_{33}^E}(-C_{13}^E S_1 - C_{23}^E S_2 + e_{33} E_3) \tag{6.28}$$

that holds for the thin plate when it is assumed to be in a state of plane stress. This constraint can be back substituted into the rows in Equation 6.27 associated with T_1, T_2, and D_3.

$$T_1 = \left(C_{11}^E - \frac{C_{13}^{E,2}}{C_{33}^E} \right) S_1 + \left(C_{12}^E - \frac{C_{13}^E C_{23}^E}{C_{33}^E} \right) S_2 - \left(e_{33} - \frac{C_{13}^E e_{33}}{C_{33}^E} \right) E_3$$

$$T_2 = \left(C_{12}^E - \frac{C_{13}^E C_{23}^E}{C_{33}^E} \right) S_1 + \left(C_{22}^E - \frac{C_{23}^{E,2}}{C_{33}^E} \right) S_2 - \left(e_{33} - \frac{C_{23}^E e_{33}}{C_{33}^E} \right) E_3$$

$$D_3 = \left(e_{31} - \frac{C_{13}^E e_{33}}{C_{33}^E} \right) S_1 + \left(e_{32} - \frac{C_{23}^E e_{33}}{C_{33}^E} \right) S_2 - \left(\epsilon_{33} - \frac{e_{33}^2}{C_{33}^E} \right) E_3 \tag{6.29}$$

We summarize the constitutive laws for thin plates in a state of plane stress by collecting these expressions in the reduced matrix equation

$$\begin{Bmatrix} T_1 \\ T_2 \\ T_6 \\ D_3 \end{Bmatrix} = \begin{bmatrix} C_{11}^P & C_{12}^P & 0 & -e_{31}^P \\ C_{12}^P & C_{22}^P & 0 & -e_{32}^P \\ 0 & 0 & C_{66}^P & 0 \\ e_{31}^P & e_{32}^P & 0 & \epsilon_{33}^P \end{bmatrix} \begin{Bmatrix} S_1 \\ S_2 \\ S_6 \\ E_3 \end{Bmatrix} \tag{6.30}$$

where the constants $C_{11}^p, C_{12}^p, C_{22}^p, e_{31}^p, e_{32}^p, \epsilon_{33}^P$ are defined in Equations 6.29. These constants can be found in many places, for example, in [5]. For an alternative representation of the plate constitutive coefficients in terms of compliances, see [14].

The explicit form of the stress resultants can now be computed by substituting the constitutive law for the plate in Equation 6.30 into the definitions in Equation 6.22. We have

$$\begin{aligned}
\mathcal{M}_1 &= \int x_3 (C_{11}^P S_1 + C_{12}^P S_2 - e_{31}^P E_3) dx_3 \\
&= -C_{11}^P I \frac{\partial^2 u_3}{\partial x_1^2} - C_{12}^P I \frac{\partial^2 u_3}{\partial x_2^2} - \int x_3 e_{31}^P E_3 dx_3 \\
\mathcal{M}_2 &= \int x_3 (C_{12}^P S_1 + C_{22}^P S_2 - e_{32}^P E_3) dx_3 \\
&= -C_{12}^P I \frac{\partial^2 u_3}{\partial x_1^2} - C_{22}^P I \frac{\partial^2 u_3}{\partial x_2^2} - \int x_3 e_{32}^P E_3 dx_3 \\
\mathcal{M}_{12} &= \int x_3 T_{12} dx_3 = -2 C_{66}^P I \frac{\partial^2 u_3}{\partial x_1 \partial x_2} \\
\mathcal{M}_{21} &= \int x_3 T_{21} dx_3 = -2 C_{66}^P I \frac{\partial^2 u_3}{\partial x_2 \partial x_1}
\end{aligned} \tag{6.31}$$

The strong form of the governing equations is obtained by substituting Equations 6.31 into Equation 6.26.

$$\begin{aligned}
\rho h \frac{\partial^2 u_3}{\partial t^2} = - \Bigg\{ & \frac{\partial^2}{\partial x_1^2} \left(C_{11}^P I \frac{\partial^2 u_3}{\partial x_1^2} \right) + \frac{\partial^2}{\partial x_2^2} \left(C_{22}^P I \frac{\partial^2 u_3}{\partial x_2^2} \right) \\
& + \frac{\partial^2}{\partial x_1^2} \left(C_{12}^P I \frac{\partial^2 u_3}{\partial x_2^2} \right) + \frac{\partial^2}{\partial x_2^2} \left(C_{12}^P I \frac{\partial^2 u_3}{\partial x_1^2} \right)
\end{aligned}$$

$$+4 \frac{\partial^2}{\partial x_1 \partial x_2} \left(C_{66}^P I \frac{\partial^2 u_3}{\partial x_1 \partial x_2} \right) \Bigg\}$$
$$- \frac{\partial^2}{\partial x_1^2} \left(\int x_3 e_{31}^P E_3 dx_3 \right) - \frac{\partial^2}{\partial x_2^2} \left(\int x_3 e_{32}^P E_3 dx_3 \right) \tag{6.32}$$

Example 6.4.1 Equations 6.32 are the final form of the equations governing the elastic plate with bonded piezoceramic patches depicted in Figure 6.6. In this example we study one particular case in which the plate has a uniform thickness and cross sectional properties. We also assume that the contribution of the piezoelectric patches to the stiffness properties of the plate are negligible and that the host or substrate material is isotropic and linearly elastic. In this case it can be shown that

$$C_{11}^P = C_{22}^P = \frac{Y}{1-\nu^2}$$
$$C_{12}^P = \frac{\nu Y}{1-\nu^2}$$
$$C_{66}^P = \frac{Y}{2(1+\nu)}$$
$$2C_{12}^P + 4C_{66}^P = \frac{2Y}{1-\nu^2}$$

where Y is the Young's modulus of the substrate material and ν is its Poisson's ratio. We let $\chi_p(x_1, x_2)$ be the characteristic function of the patch so that

$$\chi_p(x_1, x_2) = \begin{cases} 1 & (x_1, x_2) \text{ is in the patch} \\ 0 & \text{otherwise.} \end{cases}$$

Across the portion of the plate to which the piezoelectric patches are bonded, the electric field is assumed to take the form

$$E_3(t, \boldsymbol{x}) := \begin{cases} -\dfrac{V(t)}{t_p} \chi_p(x_1, x_2) & x_3 \in \left[\dfrac{t_s}{2}, \dfrac{t_s}{2} + t_p \right] \\ \dfrac{V(t)}{t_p} \chi_p(x_1, x_2) & x_3 \in \left[-\left(\dfrac{t_s}{2} + t_p \right), \dfrac{t_s}{2} \right] \\ 0 \text{ otherwise} & \end{cases}$$

We then integrate the control influence terms to find that

$$\int x_3 e_{31}^P E_3 dx_3 = -e_{31}^P (t_s + t_p) \chi_P(x_1, x_2) V(t),$$
$$\int x_3 e_{32}^P E_3 dx_3 = -e_{32}^P (t_s + t_p) \chi_P(x_1, x_2) V(t).$$

The final form of the equations governing the motion of the piezoelectrically actuated plate then become

$$\rho_m h \frac{\partial^2 u_3}{\partial t^2} = - D^P \left\{ \frac{\partial^4 u_3}{\partial x_1^4} + 2\frac{\partial^4 u_3}{\partial x_1^2 \partial x_2^2} + \frac{\partial^4 u_3}{\partial x_2^4} \right\}$$
$$+ \left\{ e_{31}^P \frac{\partial^2}{\partial x_1^2}(\chi_P(x_1, x_2)) + e_{32}^P \frac{\partial^2}{\partial x_2^2}(\chi_P(x_1, x_2)) \right\} (t_s + t_p)V(t).$$

where D^P is the plate stiffness modulus

$$D^P := \frac{2h^3 Y}{3(1 - \nu^2)}.$$

These equations are essentially identical to the model found in [2, 3].

In Example 6.1.1 it is shown that even though the electrical and mechanical field variables are coupled for linearly piezoelectric structures, it is often the case that some approximation of the electrical field can be used to obtain a satisfactory estimate of the mechanical response. This method is presented in Sections 6.2, 6.3, and 6.4. In those sections it is further assumed that the unknown displacement $u_3 = u_3(t, x)$ (where $x = x_3, x_1$, or $x = (x_1, x_2)$, respectively) is given by a single function of time and space. As explained in passing in [44] and studied in detail in [27], when the material constants are discontinuous, a useful strategy for constructing a solution $u_3(t, x)$ makes the assumption that $u(t, x)$ takes the form of a piecewise function $u_3(t, x) = \sum_{i=1}^{n} u_{3,i}(t, x)\chi_{\Omega_i}$ where $\Omega = \cup_{i=1}^{n} \Omega_i$. This strategy is explored in the following example.

Example 6.4.2 Consider the piezoelectric composite beam that is depicted in Figure 6.7 which is actuated by two surface mounted piezoelectric patches that are driven by the voltages $V_1(t)$ and $V_2(t)$. Because the figure is already complicated, the circuits driving the bottom patches are not shown. In other words, the left bottom piezoelectric patch is actuated by the voltage V_1, and the bottom right patch is actuated by the voltage V_2. We also assume in this example that the base motion $z(t) =$ constant.

Following the same general strategy that is used in Example 6.3, the governing equation for this piezoelectric composite can be written as

$$\rho_m(x_1)A(x_1)\frac{\partial^2 u_3}{\partial t^2} = - \frac{\partial^2}{\partial x_1^2}\left(C_{11}^E(x_1)I(x_1)\frac{\partial^2 u_3}{\partial x_1^2} \right)$$

(Continued)

Example 6.4.2 (Continued)

$$+\frac{\partial^2}{\partial x_1^2}\left(\frac{\kappa(x_1)e_{31}(x_1)\chi_{[a_1,b_1]}}{t_p(x_1)}\right)V_1$$
$$+\frac{\partial^2}{\partial x_1^2}\left(\frac{\kappa(x_1)e_{31}(x_1)\chi_{[a_2,b_2]}}{t_p(x_1)}\right)V_2,$$

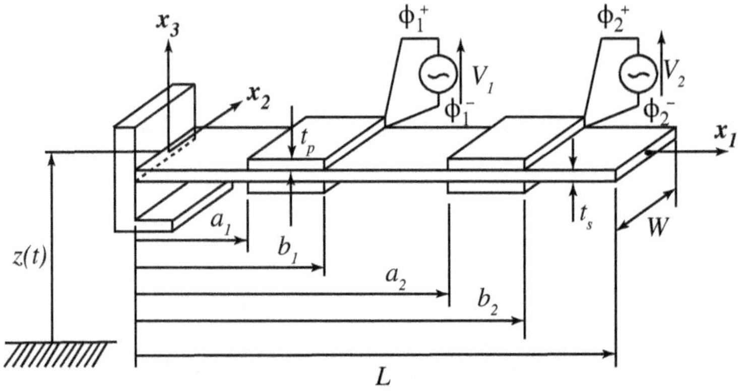

Figure 6.7 Composite piezoelectric beam actuated by two patches.

which holds for $x \in \Omega = [0, L]$. This equation, which is expressed in terms of the spatially dependent materials, is equivalent to a collection of partial differential equations that hold over subdomains Ω_i of the domain $\Omega = \cup_{i=1}^{5}$. Define the subdomains $\Omega_1 := [0, a_1], \Omega_2 := [a_1, b_1], \Omega_3 := [b_1, a_2], \Omega_4 := [a_2, b_2], \Omega_5 := [b_2, L]$. The material properties are in fact taken to be the piecewise constant functions given by

$$A(x_1) := \sum_{i=1}^{5} A_i \chi_{\Omega_i} \quad \rho_m(x_1) := \sum_{i=1}^{5} \rho_{m,i} \chi_{\Omega_i} \quad C_{11}^E(x_1) := \sum_{i=1}^{5} C_{11,i}^E \chi_{\Omega_i},$$
$$\kappa(x_1) := \sum_{i=1}^{5} \kappa_i \chi_{\Omega_i} \quad I(x_1) := \sum_{i=1}^{5} I_i \chi_{\Omega_i} \qquad e_{31}(x_1) := \sum_{i=1}^{5} e_{31,i} \chi_{\Omega_i},$$
$$t_p(x_1) := \sum_{i=1}^{5} t_{p,i} \chi_{\Omega_i}.$$

where $A_i, \rho_{m,i}, C^E_{11,i}, \kappa_i, e_{31,i}, t_{p,i}$ are constants over Ω_i for $i = 1, \ldots, 5$. We likewise assume that the displacement $u_3(t, x)$ is given by the piecewise definition

$$u_3(t,x) = \begin{cases} u_{3,1}(t,x_1) & x_1 \in \Omega_1 = [0, a_1], \\ u_{3,2}(t,x_1) & x_1 \in \Omega_2 = [a_1, b_1], \\ u_{3,3}(t,x_1) & x_1 \in \Omega_3 = [b_1, a_2], \\ u_{3,4}(t,x_1) & x_1 \in \Omega_4 = [a_2, b_2], \\ u_{3,5}(t,x_1) & x_1 \in \Omega_5 = [b_2, L]. \end{cases}$$

With these piecewise definitions of the constants and unknown displacement field, we obtain the following five evolutionary partial differential equations:

$$\rho_{m,1}A_1\frac{\partial^2 u_{3,1}}{\partial t^2} = -\frac{\partial^2}{\partial x_1^2}\left(C^E_{11,1}I_1\frac{\partial^2 u_{3,1}}{\partial x_1^2}\right) \qquad \text{for } x_1 \in \Omega_1 = [0, a_1],$$

$$\begin{aligned}\rho_{m,2}A_2\frac{\partial^2 u_{3,2}}{\partial t^2} &= -\frac{\partial^2}{\partial x_1^2}\left(C^E_{11,2}I_2\frac{\partial^2 u_{3,2}}{\partial x_1^2}\right) \\ &\quad + \frac{\partial^2}{\partial x_1^2}\left(\frac{\kappa_1 e_{31,1}\chi_{[a_1,b_1]}}{t_{p,1}}\right)V_1 \qquad \text{for } x_1 \in \Omega_2 = [a_1, b_1],\end{aligned}$$

$$\rho_{m,3}A_3\frac{\partial^2 u_{3,3}}{\partial t^2} = -\frac{\partial^2}{\partial x_1^2}\left(C^E_{11,3}I_3\frac{\partial^2 u_{3,3}}{\partial x_1^2}\right) \qquad \text{for } x_1 \in \Omega_3 = [b_1, a_2],$$

$$\begin{aligned}\rho_{m,4}A_4\frac{\partial^2 u_{3,4}}{\partial t^2} &= -\frac{\partial^2}{\partial x_1^2}\left(C^E_{11,4}I_4\frac{\partial^2 u_{3,4}}{\partial x_1^2}\right) \\ &\quad + \frac{\partial^2}{\partial x_1^2}\left(\frac{\kappa_4 e_{31,4}\chi_{[a_2,b_2]}}{t_{p,4}}\right)V_2 \qquad \text{for } x_1 \in \Omega_4 = [a_2, b_2],\end{aligned}$$

$$\rho_{m,5}A_5\frac{\partial^2 u_{3,5}}{\partial t^2} = -\frac{\partial^2}{\partial x_1^2}\left(C^E_{11,5}I_5\frac{\partial^2 u_{3,5}}{\partial x_1^2}\right) \qquad \text{for } x_1 \in \Omega_5 = [b_2, L].$$

Each of the five problems for the unknowns $u_{3,i}(t, x_1)$ for $i = 1, \ldots, 5$ is subject to its own collection of boundary conditions as listed in the table below.

(Continued)

Example 6.4.2 (Continued)

Table 6.1 Boundary conditions for piezoelectric composite beam driven by two voltage sources.

Domains	Original BC	Compatibility BC
Ω_1	$u_{3,1}(t,0)=0$ $\frac{\partial u_{3,1}}{\partial x_1}(t,0)=0$	
Ω_1 and Ω_2		$u_{3,1}(t,a_1)=u_{3,2}(t,a_1)$ $\frac{\partial u_{3,1}}{\partial x_1}(t,a_1)=\frac{\partial u_{3,2}}{\partial x_1}(t,a_1)$ $\mathcal{S}_1(t,a_1)=\mathcal{S}_2(t,a_1)$ $\mathcal{M}_1(t,a_1)=\mathcal{M}_2(t,a_1)$
Ω_2 and Ω_3		$u_{3,2}(t,b_1)=u_{3,3}(t,b_1)$ $\frac{\partial u_{3,2}}{\partial x_1}(t,b_1)=\frac{\partial u_{3,3}}{\partial x_1}(t,b_1)$ $\mathcal{S}_2(t,b_1)=\mathcal{S}_3(t,b_1)$ $\mathcal{M}_2(t,b_1)=\mathcal{M}_3(t,b_1)$
Ω_3 and Ω_4		$u_{3,3}(t,a_2)=u_{3,4}(t,a_2)$ $\frac{\partial u_{3,3}}{\partial x_1}(t,a_2)=\frac{\partial u_{3,4}}{\partial x_1}(t,a_2)$ $\mathcal{S}_3(t,a_2)=\mathcal{S}_4(t,a_2)$ $\mathcal{M}_3(t,a_2)=\mathcal{M}_4(t,a_2)$
Ω_4 and Ω_5		$u_{3,4}(t,b_2)=u_{3,5}(t,b_2)$ $\frac{\partial u_{3,4}}{\partial x_1}(t,b_2)=\frac{\partial u_{3,5}}{\partial x_1}(t,b_2)$ $\mathcal{S}_4(t,b_2)=\mathcal{S}_5(t,b_2)$ $\mathcal{M}_4(t,b_2)=\mathcal{M}_5(t,b_2)$
Ω_5	$\mathcal{S}_5(t,L)=0$ $\mathcal{M}_5(t,L)=0$	

In summary, we obtain five evolutionary partial differential equations, each of which is written in terms of a fourth order spatial differential operator, and there consequently are four boundary conditions for each equation, making a total of 20 boundary conditions for the coupled system of equations. There are two initial conditions $\overline{u}_{0,3,i}(x_1)$ and $\overline{v}_{0,3,i}(x_1)$ for $u_{3,i}(0,x_1)$ and $\frac{\partial u_{3,i}}{\partial t}(0,x_1)$, respectively, that hold for all $x_1 \in \Omega_i$ for the equations $i = 1, \ldots, 5$. We return to this example in Chapters 7 and 8. The approximate solution of these governing equations is discussed in detail in [27].

6.5 Problems

Problems 6.5.1 Derive the governing equations of motion for the axial piezoelectric specimen with the rigidly attached tip mass M (Figure 6.8).

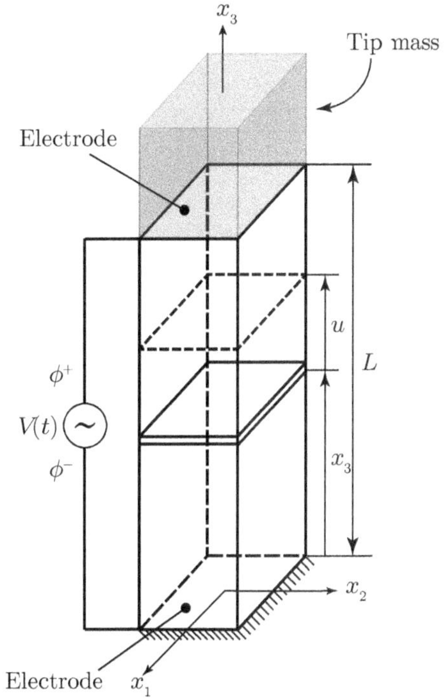

Figure 6.8 Axial piezoelectric specimen with tip mass.

Problems 6.5.2 Derive the governing equations of motion for the axial piezoelectric specimen which has a prescribed base motion given by $z(t)$ (Figure 6.9).

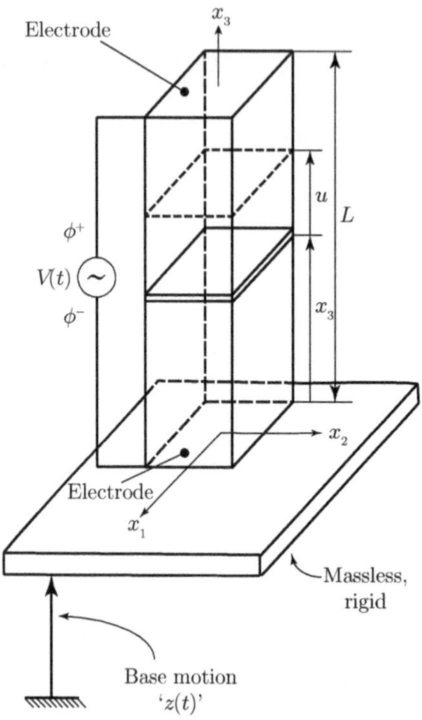

Figure 6.9 Axial piezoelectric specimen with prescribed base motion $z(t)$.

Problems 6.5.3 Derive the governing equations of motion for the piezoelectric composite beam in bending which has a rigidly attached tip mass M (Figure 6.10).

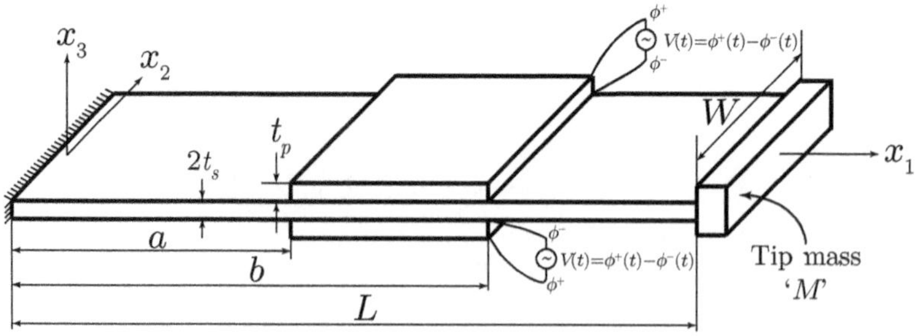

Figure 6.10 Piezoelectric composite beam specimen with rigidly attached tip mass M.

Problems 6.5.4 Derive the governing equations of motion for the piezoelectric composite beam in bending which has a prescribed base motion $z(t)$ (Figure 6.11).

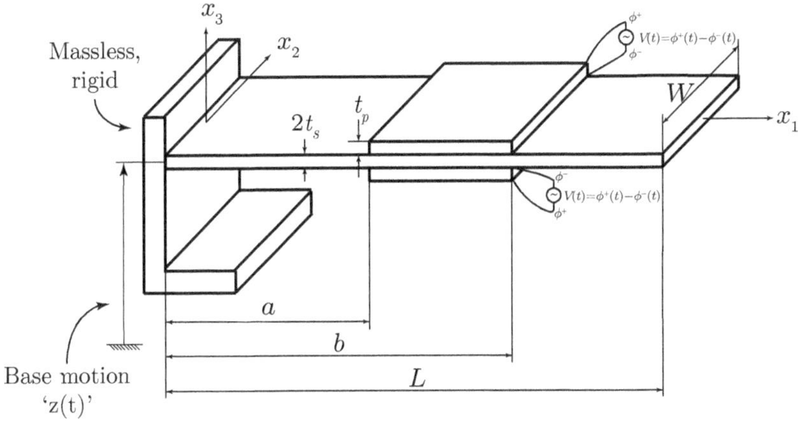

Figure 6.11 Piezoelectric composite beam specimen prescribed base motion $z(t)$.

7

Variational Methods

The equations of motion for the linearly piezoelectric systems studied in Chapters 5 and 6 are derived by carrying out force and moment summations and applying Newton's laws. For many structural systems it can be advantageous to formulate the governing equations using variational methods. These techniques are one of the standard tools used in the study of the dynamics or vibrations of linearly elastic structural systems, and this chapter discusses their generalization for linearly piezoelectric systems. We begin with a review of the underlying theory of variational methods in Section 7.1. We next review Hamilton's principle as it is applied to linearly elastic systems in Section 7.2. Hamilton's principle for linear piezoelectricity is subsequently presented in Section 7.3.

7.1 A Review of Variational Calculus

One of the first applications of classical differential calculus that a student encounters is the characterization of extrema of real valued functions. We learn that the extrema occur at those points for which the derivative of the function is equal to zero. In many applications in mechanics, and specifically in the study of linearly piezoelectric materials, we are interested in studying the extrema of functionals instead of real valued functions. A functional J is a mapping from some abstract (vector) space to the real line, that is, $J : \mathcal{F} \to \mathbb{R}$. To understand the structure of the particular functionals that we study in this chapter, we must first define generalized coordinates, configuration space Q, and trajectories $\mathcal{F}$ in configuration space Q. A set of time-varying parameters $\left\{q_1(t)\ q_2(t)\ ,\cdots\ ,\ q_n(t)\right\}_{t\in\mathbb{R}^+}$ constitute a set of generalized coordinates for a mechanical system if the inertial position $\boldsymbol{r}_{I,p}(t)$ of every point p in any possible configuration of the mechanical system at any time t can be expressed uniquely as $\boldsymbol{r}_{I,p}(t) = \boldsymbol{r}_{I,p}(t, q_1(t), \cdots, q_n(t))$. One implication of this definition is that the generalized coordinates are independent: there are no constraints that couple the generalized coordinates. The number of

Vibrations of Linear Piezostructures, First Edition. Andrew J. Kurdila and Pablo A. Tarazaga.

This Work is a co-publication between John Wiley & Sons Ltd and ASME Press.

generalized coordinates n is known as the number of degrees of freedom of the mechanical system. Configuration space is the set of all the admissible values of the generalized coordinates as time is varied. A trajectory or motion in $\mathcal{F}$ is the curve $t \mapsto \{q_1(t) \cdots q_n(t)\}$ traced out in configuration space Q. With these definitions in place, the abstract definition of a functional as a mapping $J : \mathcal{F} \to \mathbb{R}$ can be understood by considering two specific examples of particular importance in this text.

Example 7.1.1 We consider a linearly elastic continuum that occupies the volume Ω. The configuration of the continuum is determined by the time histories, or trajectories, of n generalized coordinates $q(t) := \{q_1(t), \dots, q_n(t)\}$, for $t \in \mathbb{R}^+$. Let Q be the configuration space generated by all such admissible trajectories. We assume that the kinetic energy $\mathcal{T}$ depends on the generalized coordinates and their derivatives, and the strain energy is given by $\mathcal{V}(q(t)) = \int_\Omega \mathcal{U}_0(q(t))dv$ where $\mathcal{U}_0$ is the strain energy density. We define the action integral $\mathcal{A}$ as

$$\mathcal{A}(q) := \int_{t_0}^{t_1} (\mathcal{T}(q(t), \dot{q}(t), t) - \mathcal{V}(q(t)))dt.$$

The action integral $\mathcal{A} : \mathcal{F} \to \mathbb{R}$ is a functional.

Example 7.1.2 Suppose that we consider a linearly piezoelectric continuum that occupies the volume Ω. As in the last example, suppose that the state of the body is determined by the time histories of n generalized coordinates $q(t) := \{q_1(t), \dots, q_n(t)\}$ for $t \in \mathbb{R}^+$, and let Q be the configuration space generated by all such admissible trajectories of these coordinates. Suppose further that we define the electric enthalpy $\mathcal{V}_\mathcal{H}(q(t)) = \int_\Omega \mathcal{H}(q(t))dt$ where $\mathcal{H}$ is the electric enthalpy density. We define the piezoelectric action integral as

$$\mathcal{A}_\mathcal{H}(q) := \int_{t_0}^{t_1} (\mathcal{T}(q(t), \dot{q}(t), t) - \mathcal{V}_\mathcal{H}(q(t)))dt.$$

The piezoelectric action integral is a functional $\mathcal{A}_\mathcal{H} : Q \to \mathbb{R}$.

It is important to note in Examples 7.1.1 and 7.1.2 that the functionals $\mathcal{A}$ and $\mathcal{A}_\mathcal{H}$ act on trajectories, or time histories, that are contained in the vector space $\mathcal{F}$. These functionals are not examples of real valued functions defined on subsets of $\mathbb{R}^n$, and the classical definition of a derivative does not apply.

Variational calculus studies the extrema of functionals, such as the action integrals introduced in Examples 7.1.1 and 7.1.2. We will see that if we define the derivative of a functional appropriately, the extrema of the functional are characterized by the condition that the derivative is equal to zero. The Gateaux derivative

$DJ(q,p)$ of a functional J is perhaps the simplest definition of the derivative of the functional that can be used to study its extrema.

Definition 7.1.1 The Gateaux derivative $DJ(q,p)$ of a functional J at $q \in \mathcal{F}$ in the direction $p \in \mathcal{F}$ is defined to be

$$DJ(q,p) := \lim_{\epsilon \to 0} \frac{J(q+\epsilon p) - J(q)}{\epsilon}.$$

It is possible to calculate the Gateaux derivative of a functional J directly from the Definition 7.1.1. Theorem 7.1.1 provides an alternative means of calculating the Gateaux derivative that can be convenient in many applications.

Theorem 7.1.1 If the functional J is Gateaux differentiable, we have

$$DJ(q,p) = \frac{d}{d\epsilon} J(q+\epsilon p)\Big|_{\epsilon=0}$$

It can be shown [25] that if a Gateaux differentiable functional $J(q)$ has a local extrema at $q \in \mathcal{F}$, then $DJ(q,p) = 0$ for all admissible choices of directions $p \in \mathcal{F}$. The calculation of the Gateaux derivative of a functional has proven to be an effective way of deriving the governing equations in a wide range of applications. We show how the Gateaux derivative can be calculated by considering a classical example, the derivation of Lagrange's equations.

Example 7.1.3 In this example we continue our study of the action functional introduced in Example 7.1.1 and calculate its Gateaux derivative. Using Theorem 7.1.1 we can write

$$\begin{aligned}
D\mathcal{A}(q,p) &= \frac{d}{d\epsilon} \mathcal{A}(q+\epsilon p)|_{\epsilon=0} \\
&= \int_{t_0}^{t_1} \left(\frac{\partial \mathcal{T}(q(t)+\epsilon p(t), \dot{q}(t)+\epsilon \dot{p}(t), t)}{\partial (q_k + \epsilon p_k)} p_k(t) \right. \\
&\quad \left. + \frac{\partial \mathcal{T}(q(t)+\epsilon p(t), \dot{q}(t)+\epsilon \dot{p}(t), t)}{\partial (\dot{q}_k + \epsilon \dot{p}_k)} \dot{p}_k(t) - \frac{\partial \mathcal{V}(q(t)+\epsilon p(t))}{\partial (q_k + \epsilon p_k)} p_k(t) \right) dt \Bigg|_{\epsilon=0} \\
&= \int_{t_0}^{t_1} \left(\frac{\partial \mathcal{T}(q(t), \dot{q}(t), t)}{\partial q_k} p_k(t) + \frac{\partial \mathcal{T}(q(t), \dot{q}(t), t)}{\partial \dot{q}_k} \dot{p}_k(t) - \frac{\partial \mathcal{V}(q(t))}{\partial q_k} p_k(t) \right) dt \\
&= \int_{t_0}^{t_1} \left(\frac{\partial \mathcal{T}(q(t), \dot{q}(t), t)}{\partial q_k} - \frac{d}{dt} \left(\frac{\partial \mathcal{T}(q(t), \dot{q}(t), t)}{\partial \dot{q}_k} \right) - \frac{\partial \mathcal{V}(q(t))}{\partial q_k} \right) p_k(t) dt \\
&\quad + \frac{\partial \mathcal{T}(q(t), \dot{q}(t), t)}{\partial \dot{q}_k} p_k(t) \Bigg|_{t_0}^{t_1},
\end{aligned} \tag{7.1}$$

(Continued)

Example 7.1.3 (Continued)

with the summation over $k = 1, \dots, n$. The final result follows when we integrate by parts to eliminate $\dot{p}_k$ for $k = 1, \dots, n$. If the action functional has a local extrema, it follows that Equation 7.1 must hold for all admissible directions $p := \{p_1(t), \dots, p_n(t)\}^T$. Since we require that the varied trajectory $q(t) + \epsilon p(t)$ satisfies the same initial and final conditions as $q(t)$, the admissible directions p must be such that that $p_k(t_0) = p_k(t_1)$ for $k = 1, \dots, n$. We conclude that Lagrange's equations

$$\frac{\partial \mathcal{T}(q(t), \dot{q}(t), t)}{\partial q_k} - \frac{d}{dt}\left(\frac{\partial \mathcal{T}(q(t), \dot{q}(t), t)}{\partial \dot{q}_k}\right) - \frac{\partial \mathcal{V}(q(t))}{\partial q_k} = 0$$

hold for $k = 1, \dots, n$ when the action functional is stationary.

Even with the use of Theorem 7.1.1, Example 7.1.3 illustrates that the calculation of the Gateaux derivative of a nontrivial functional using the definition can be lengthy. Some authors [11, 17] introduce the (virtual) variation operator $\delta(\cdot)$ to simplify the calculation of Gateaux derivatives such as those shown in Example 7.1.3. In summary, the virtual variation operator $\delta(\cdot)$ is assumed to satisfy the following conditions:

(V1) The variation $\delta(\cdot)$obeys the same rules as the differential operator $d(\cdot)$.
(V2) The variation of the independent variable t is zero, $\delta t \equiv 0$.
(V3) The variation commutes with differentiation, $\delta(\frac{d}{dt}(\cdot)) = \frac{d}{dt}(\delta(\cdot))$ and $\delta(\frac{d}{dx_i}(\cdot)) = \frac{d}{dx_i}(\delta(\cdot))$.
(V4) The variation $\delta q(t)$ is equal to zero at t_0 and t_1, $\delta q(t_0) = \delta q(t_1) = 0$. (7.2)

The following example shows how the application of the variation operator can simplify the calculation of the Gateaux derivative.

Example 7.1.4 We again return to Example 7.1.3. We apply the variational operator $\delta(\cdot)$ to find

$$\delta(\mathcal{T} - \mathcal{V})dt = \int_{t_0}^{t_1} \left(\frac{\partial \mathcal{T}}{\partial q_k}\delta q_k + \frac{\partial \mathcal{T}}{\partial \dot{q}_k}\delta(\dot{q}_k) - \frac{\partial \mathcal{V}}{\partial q_k}\delta q_k\right) dt \tag{7.3}$$

$$= \int_{t_0}^{t_1} \left\{\frac{\partial \mathcal{T}}{\partial q_k} - \frac{d}{dt}\left(\frac{\partial \mathcal{T}}{\partial \dot{q}_k}\right) - \frac{\partial \mathcal{V}}{\partial q_k}\right\} \delta q_k dt - \sum_{k=1}^{n} \frac{\partial \mathcal{T}}{\partial \dot{q}_k}\delta q_k\Bigg|_{t_0}^{t_1} \tag{7.4}$$

with the summation in $k = 1, \dots, n$. The first line in Equation 7.3 follows from (V1) and (V2). That is, the variation $\delta(\cdot)$ operates like the differential operator $d(\cdot)$ and therefore obeys a chain rule. The second line in Equation 7.4 follows from (V3) and (V4) when we integrate by parts. This result should be compared to the answer obtained in Example 7.1.3. The resulting governing equations are the same when we identify the variations δq_k and the directions p_k.

7.2 Hamilton's Principle

Hamilton's Principle is used to determine the equations of motion of a mechanical system from the stationarity of the action integral.

Theorem 7.2.1 ***(Hamilton's Principle)*** Of all the possible motions of a conservative mechanical system, the true motion is such that it renders the action integral $\mathcal{A}$ stationary. The stationarity condition for the action integral $\mathcal{A}$ of a conservative system is expressed as the requirement that

$$\delta \int_{t_0}^{t_1} (\mathcal{T} - \mathcal{V})dt = 0$$

for all admissible variations δq of a trajectory q in configuration space.

This form of Hamilton's principle emphasizes that it is naturally interpreted as a statement of variational calculus. The techniques reviewed in Section 7.1 for the study of the extrema of functionals can therefore be applied. In particular, we can compute the Gateaux derivative of the action integral to find the equations of motion for the system. If the system is subject to nonconservative forces, we apply Hamilton's Extended Principle.

Theorem 7.2.2 ***(Hamilton's Extended Principle)*** For a nonconservative mechanical system, we have

$$\delta \int_{t_0}^{t_1} (\mathcal{T} - \mathcal{V})dt + \int_{t_0}^{t_1} \delta W dt = 0$$

for all admissible variations δq of the true motion q in configuration space where δW is the virtual work performed by the nonconservative forces.

The virtual work δW contains contributions of prescribed, nonconservative, mechanical loads. Suppose that the body is subject to nonconservative point forces $\boldsymbol{f}_i$ for $i = 1, \ldots, n_f$ and applied tractions $\overline{\boldsymbol{\tau}}$ on $\partial\Omega_T$. The virtual work δW of the applied mechanical loads is written as

$$\delta W = \boldsymbol{f}_i \cdot \delta \boldsymbol{u}_i + \int_{\partial\Omega_T} \overline{\boldsymbol{\tau}} \cdot \delta \boldsymbol{u} da$$

where $\delta \boldsymbol{u}_i$ is the virtual displacement of the point of application of the nonconservative force $\boldsymbol{f}_i$ for $i = 1, \ldots n_f$.

We show how this principle is applied in two common cases, summarized in Sections 7.2.1 and 7.2.2.

7.2.1 Uniaxial Rod

If we set the piezoelectric constant $e_{33} = 0$ in the governing Equation 6.11 for the linearly piezoelectric axial structural element, we see that the equation of motion for the free vibration of a linearly elastic uniaxial structural element is given by

$$\rho_m A \frac{\partial^2 u_3}{\partial^2 t} = \frac{\partial}{\partial x_3}\left(AC_{33}\frac{\partial u_3}{\partial x_3}\right). \tag{7.5}$$

We can also obtain this result directly by invoking Hamilton's principle and solving the resulting problem of variational calculus. The kinetic energy of the axial rod is given by

$$\mathcal{T} = \frac{1}{2}\int_0^L \rho_m A \left(\frac{\partial u_3}{\partial t}\right)^2 dx_3,$$

and the general form of the strain energy density $\mathcal{U}_0$ for a linearly elastic material has been derived in Section 3.3 as

$$\mathcal{U}_0 = \frac{1}{2}T_{ij}S_{ij} \approx \frac{1}{2}C_{33}S_{33}^2.$$

We integrate the strain energy density $\mathcal{U}_0$ over the domain Ω to obtain the strain energy $\mathcal{V}$, which for this particular geometry takes the specific form

$$\begin{aligned}\mathcal{V} = \int_\Omega \mathcal{U}_0 dv &= \int_0^L \iint \frac{1}{2}C_{33}\left(\frac{\partial u_3}{\partial x_3}\right)^2 dx_1 dx_2 dx_3,\\ &= \frac{1}{2}\int_0^L AC_{33}\left(\frac{\partial u_3}{\partial x_3}\right)^2 dx_3.\end{aligned}$$

We calculate the Gateaux derivative by using the variational operator $\delta(\cdot)$ introduced in Section 7.1:

$$\delta \int_{t_0}^{t_1} (\mathcal{T} - \mathcal{V})dt = \int_{t_0}^{t_1}\int_0^L \left\{\rho_m A \frac{\partial u_3}{\partial t}\delta\left(\frac{\partial u_3}{\partial t}\right) - AC_{33}\frac{\partial u_3}{\partial x_3}\delta\left(\frac{\partial u_3}{\partial x_3}\right)\right\} dx_3 dt, \tag{7.6}$$

$$= \int_{t_0}^{t_1} \int_0^L \left\{ -\rho_m A \frac{\partial^2 u_3}{\partial t^2} + \frac{\partial}{\partial x_3} \left(AC_{33} \frac{\partial u_3}{\partial x_3} \right) \right\} \delta u_3 dx_3 dt, \tag{7.7}$$

$$+ \int_{t_0}^{t_1} \left\{ -AC_{33} \frac{\partial u_3}{\partial x_3} \delta u_3 \Big|_0^L \right\} dt. \tag{7.8}$$

The first line in Equation 7.6 follows from the fact that the variational operator $\delta(\cdot)$ obeys a chain rule like that for the differential operator $d(\cdot)$. The second and third lines in Equations 7.7 and 7.8 is obtained by using the fact that the variational operator $\delta(\cdot)$ commutes with differentiation, and then integrating by parts to eliminate the derivatives that act on δu_3. Since the equations 7.7 and 7.8 must hold true for all choices of admissible variations δu_3, we see that a solution $u_3(t, x_3)$ satisfies the governing Equation 7.5,

$$\rho_m A \frac{\partial^2 u_3}{\partial t^2} = \frac{\partial}{\partial x_3} \left(AC_{33} \frac{\partial u_3}{\partial x_3} \right),$$

subject to the variational boundary conditions

$$AC_{33} \frac{\partial u_3}{\partial x_3} \delta u_3 \Big|_0^L = 0,$$

and subject to the the initial conditions

$$u_3(0, x_3) = \overline{u}_3(x_3),$$

$$\frac{\partial u_3}{\partial t}(0, x_3) = \overline{v}_3(x_3),$$

for all $x_3 \in [0, L]$ when $t = 0$.

The variational boundary conditions in Equation 7.2.1 are a succinct way of summarizing a variety of types of physical conditions. If the displacement u_3 is fixed to a constant value at either $x_3 = 0$ or $x_3 = L$, the corresponding variations $\delta u_3(t, 0)$ or $\delta u_3(t, L)$, respectively, must be zero. On the other hand, if $u_3(t, x_3)$ is free to vary at either $x_3 = 0$ or $x_3 = L$, then the variations $\delta u_3(t, 0)$ or $\delta u_3(t, L)$, respectively, are arbitrary. Suppose for example that the lower end of the axial element is fixed so that $u(t, 0) = 0$ and that the end at $x_3 = L$ is free. Then the variational boundary conditions are written as

$$\begin{aligned} 0 &= AC_{33} \frac{\partial u_3}{\partial x_3} \delta u_3 \Big|_0^L, \\ &= AC_{33} \frac{\partial u_3}{\partial x_3}(t, L) \underbrace{\delta u_3(t, L)}_{\text{arbitrary}} - AC_{33} \frac{\partial u_3}{\partial x_3}(t, 0) \underbrace{\delta u_3(t, 0)}_{=0}. \end{aligned}$$

We conclude, since $\delta u_3(t, L)$ is arbitrary, that $AC_{33} \frac{\partial u_3}{\partial x_3}(t, L) = AT_{33}(t, L) = 0$. In other words, the external load on the top of the structural element is zero.

7.2.2 Bernoulli–Euler Beam

Another popular example of the application of Hamilton's principle that is used to illustrate the application of variational methods in vibrations or structural dynamics considers a linearly elastic Bernoulli–Euler beam. [11, 29] If we set the piezoelectric constant $e_{31} = 0$ in Equation 6.20, we see that the equation governing the motion of a linearly elastic composite beam takes the form

$$\rho_m A \frac{\partial^2 u_3}{\partial t^2} = -\frac{\partial}{\partial x_1^2}\left(C_{11}^E I \frac{\partial^2 u_3}{\partial x_1^2} \right).$$

This equation can be derived using Hamilton's principle by noting that the strain energy density of a linearly elastic beam can be approximated as

$$\mathcal{U}_0 = \frac{1}{2} T_{ij} S_{ij} \approx \frac{1}{2} C_{11} S_{11}^2 = \frac{1}{2} C_{11} \left(-x_3 \frac{\partial^2 u_3}{\partial x_1^2} \right)^2,$$

and therefore the strain energy for the entire beam is given by

$$\begin{aligned} \mathcal{V} &= \int_0^L \iint \frac{1}{2} C_{11} x_3^2 \left(\frac{\partial^2 u_3}{\partial x_1^2} \right)^2 dx_2 dx_3 dx_1 \\ &= \frac{1}{2} \int_0^L C_{11} I \left(\frac{\partial^2 u_3}{\partial x_1^2} \right)^2 dx_1. \end{aligned}$$

Hamilton's principle requires that the variation of the action integral is equal to zero for all admissible variations, which results in the following variational equations:

$$\delta \int_{t_0}^{t_1} (\mathcal{T} - \mathcal{V}) dt = \int_{t_0}^{t_1} \int_0^L \left\{ \rho_m A \frac{\partial u_3}{\partial t} \delta \left(\frac{\partial u_3}{\partial t} \right) - C_{11} I \frac{\partial^2 u_3}{\partial x_1^2} \delta \left(\frac{\partial^2 u_3}{\partial x_1^2} \right) \right\} dx_1 dt \tag{7.9}$$

$$= \int_{t_0}^{t_1} \int_0^L \left\{ -\rho_m A \frac{\partial^2 u_3}{\partial t^2} - \frac{\partial^2}{\partial x_1^2} \left(C_{11} I \frac{\partial^2 u_3}{\partial x_1^2} \right) \right\} \delta u_3 dx_1 dt \tag{7.10}$$

$$+ \int_{t_0}^{t_1} \left\{ -C_{11} I \frac{\partial^2 u_3}{\partial x_1^2} \delta \left(\frac{\partial u_3}{\partial x_1} \right) + \frac{\partial}{\partial x_1} \left(C_{11} I \frac{\partial^2 u_3}{\partial x_1^2} \right) \delta u_3 \right\} \Bigg|_0^L dt \tag{7.11}$$

The first line in Equation 7.9 results from the fact that the variation operator $\delta(\cdot)$ obeys the chain rule. The second and third lines in Equations 7.10 and 7.11 are obtained by using the fact that the variation operator commutes with differentiation and subsequently integrating by parts to remove any derivatives on δu_3. Since the variational equation in Equations 7.10 and 7.11 must hold for any admissible variation δu_3, we see that the displacement $u_3(t, x_1)$ must satisfy the equation

$$\rho_m A \frac{\partial^2 u_3}{\partial t^2} = -\frac{\partial^2}{\partial x_1^2}\left(C_{11} I \frac{\partial^2 u_3}{\partial x_1^2} \right)$$

for all $(t, x_1) \in \mathbb{R}^+ \times [0, L]$, subject to the variational boundary conditions

$$C_{11} I \frac{\partial^2 u_3}{\partial x_1^2} \delta\left(\frac{\partial u_3}{\partial x_1} \right)\bigg|_0^L = 0,$$

$$\frac{\partial}{\partial x_1}\left(C_{11} I \frac{\partial^2 u_3}{\partial x_1^2} \right) \delta u_3 \bigg|_0^L = 0,$$

and subject to the initial conditions

$$u_3(0, x_1) = \overline{u}_3(x_1),$$

$$\frac{\partial u_3}{\partial t}(0, x_1) = \overline{v}_3(x_1).$$

7.3 Hamilton's Principle for Piezoelectricity

Sections 7.2, 7.2.1, and 7.2.2 have shown that Hamilton's Principle provides an effective alternative to Newton's formulation for deriving the equations of motion of linearly elastic continua. Hamilton's principle for linearly piezoelectric bodies can be stated in much the same manner as that for linearly elastic bodies.

Theorem 7.3.1 ***(Hamilton's Principle for Linearly Piezoelectric Bodies)*** Let Ω be a linearly piezoelectric body that constitutes a conservative system, and denote its electric enthalpy as $\mathcal{V}_\mathcal{H} := \int_\Omega \mathcal{H} dv$. Of all the possible motions of the body, the true motion renders the piezoelectric action integral $\mathcal{A}_\mathcal{H} := \mathcal{T} - \mathcal{V}_\mathcal{H}$ stationary. The stationarity condition is expressed as the requirement that

$$0 = \delta \int_{t_0}^{t_1} (\mathcal{T} - \mathcal{V}_\mathcal{H}) dt \tag{7.12}$$

for all admissible variations of the true motion. For a nonconservative, linearly piezoelectric body, the extended Hamilton's Principle holds. It states that

$$0 = \delta \int_{t_0}^{t_1} (\mathcal{T} - \mathcal{V}_\mathcal{H}) dt + \int_{t_0}^{t_1} \delta W dt \tag{7.13}$$

for all admissible variations of the motion. In this equation δW is the virtual work of the nonconservative mechanical and electrical loads.

When we refer to the admissible variations of the motion in the above theorem we refer to the variations of the path in the electromechanical configuration space that are induced by variations in the electrical and mechanical system variables. In this text it is common that the state variables will include a displacement u and a voltages V, so that variations δu and δV induce possible trajectories or motions in the electromechanical configuration space. See [12] or [38] for discussions of Hamilton's Principle for piezoelectric or electromechanical systems in greater detail.

The computation of the nonconservative virtual work is more complicated in this case in contrast to linearly elastic systems described in Theorem 7.2.1. The virtual work δW contains contributions of prescribed mechanical loads and electrical loads. Suppose that the body is subject to nonconservative point forces $\boldsymbol{f}_i$ for $i = 1, \dots, n_f$, applied tractions $\overline{\tau}$ on $\partial\Omega_T$, prescribed currents i_k for $k = 1, \dots, n_I$, and a prescribed surface charge distributions $\overline{\sigma}$ on $\partial\Omega_D$. The virtual work δW of these applied mechanical and electrical loads is written as

$$\begin{aligned}\delta W &= \boldsymbol{f}_i \cdot \delta \boldsymbol{u}_i + \int_{\partial\Omega_T} \overline{\tau} \cdot \delta \boldsymbol{u} da \\ &\quad + i_k \delta \lambda_k - \int_{\partial\Omega_D} \overline{\sigma} \delta \phi da\end{aligned}$$

where $\delta \boldsymbol{u}_i$ is the virtual displacement of the point of application of the nonconservative force $\boldsymbol{f}_i$ for $i = 1, \dots n_f$, $\delta\lambda_k$ is the variation of the flux linkage λ_k for $k = 1, \dots, n_I$, and $\delta\phi$ is the variation of the scalar potential ϕ. It is common practice [12, 14, 38, 42] to express the virtual work of the nonconservative electrical loads in terms of the flux linkage λ_k, whose time derivative is equal to the voltage V_k,

$$\dot{\lambda}_k := V_k,$$

for $k = 1, \dots, I$. Alternative forms of the virtual work due to electrical loads can be obtained via integration by parts. We use the identity $\dot{q}_k := i_k$, the commutation relationship $\delta \frac{d}{dt}(\cdot) = \frac{d}{dt}(\delta \cdot)$, and the fact that the variations $\delta\lambda_k$ vanish at t_0 and t_1, to obtain

$$\begin{aligned}\int_{t_0}^{t_1} i_k \delta \lambda_k dt &= \int_{t_0}^{t_1} \dot{q}_k \delta \lambda_k dt, \\ &= q_k \delta \lambda_k \big|_{t_0}^{t_1} - \int_{t_0}^{t_1} q_k \frac{d}{dt}(\delta \lambda_k) dt, \\ &= -\int_{t_0}^{t_1} q_k \delta \dot{\lambda}_k dt = -\int_{t_0}^{t_1} q_k \delta V_k dt.\end{aligned}$$

Theorem 7.3.1 is applicable to a single piezoelectric, perhaps composite, body. We next generalize a form of Hamilton's Principle that is applicable to electromechanical systems that include piezoelectric bodies and discrete ideal electrical components such as inductors, capacitors, and resistors.

Theorem 7.3.2 ***(Hamilton's Principle for Electromechanical Systems)*** Let the electromechanical system under consideration consist of mechanical components, piezoelectric bodies, ideal discrete capacitors C_i for $i = 1, \dots, n_C$, ideal discrete inductors L_i for $i = 1, \dots, n_L$, or ideal resistors. Define the electromechanical potential to be

$$\mathcal{V}_{em} := \mathcal{V}_{\mathcal{H}} + \mathcal{V}_m - \mathcal{V}_e,$$

with the magnetic and electric potentials defined as

$$\mathcal{V}_m := \frac{1}{2}\frac{1}{L_i}\lambda_i^2,$$
$$\mathcal{V}_e := \frac{1}{2}C_j V_j^2,$$

respectively, and the summation convention is over $i = 1, \dots, n_L$ and $j = 1, \dots, n_C$. Of all the possible motions of a conservative electromechanical system, the true motion renders the action integral $\mathcal{A}_{em} := \mathcal{T} - \mathcal{V}_{em}$ stationary. The stationarity condition for a conservative system is expressed as the requirement that

$$0 = \delta \int_{t_0}^{t_1} (\mathcal{T} - \mathcal{V}_{em})dt \tag{7.14}$$

for all admissible variations of the true motion. For a nonconservative electromechanical system, the extended Hamilton's Principle holds. It states that

$$0 = \delta \int_{t_0}^{t_1} (\mathcal{T} - \mathcal{V}_{em})dt + \int_{t_0}^{t_1} \delta W dt \tag{7.15}$$

for all admissible variations of the motion. In this equation δW is the virtual work of the nonconservative mechanical and electrical loads.

We will now show that a motion that satisfies the variational equations in Hamilton's principle above a does indeed satisfy the governing initial-boundary value problem for linearly piezoelectric continua summarized in Equation 5.6 in Section 5.2.2. The explicit form of Hamilton's principle expressed in terms of the displacements u_i of the linearly piezoelectric continuum is given by

$$\begin{aligned}\delta \int_{t_0}^{t_1} (\mathcal{T} - \mathcal{V}_{\mathcal{H}})dt &+ \int_{t_0}^{t_1} \delta W dt \\ &= \delta \int_{t_0}^{t_1} \left\{ \frac{1}{2}\int_{\Omega} \rho_m \frac{\partial u_i}{\partial t}\frac{\partial u_i}{\partial t} dv - \int_{\Omega} \mathcal{H}(S_{ij}, E_k) dv \right\} dt \\ &+ \int_{t_0}^{t_1} \int_{\partial\Omega} (\overline{\tau}_i \delta u_i - \overline{\sigma}\delta\phi) da dt, \end{aligned} \tag{7.16}$$

where $\overline{\tau}_i$ is the externally applied surface traction and $\overline{\sigma}$ is an externally enforced surface charge distribution. The variation operator $\delta(\cdot)$ satisfies the chain rule by virtue of property (V1) in Equation 7.2 and is computed to be

$$\delta \int_{t_0}^{t_1} (\mathcal{T} - \mathcal{V}_{\mathcal{H}})dt = \int_{\Omega} \left\{ \rho_m \frac{\partial u_i}{\partial t} \delta u_i \Big|_{t_0}^{t_1} - \int_{t_0}^{t_1} \rho_m \frac{\partial^2 u_i}{\partial t^2} \delta u_i dt \right\} dv$$
$$- \int_{t_0}^{t_1} \int_{\Omega} \left\{ \frac{\partial \mathcal{H}}{\partial S_{ij}} \delta S_{ij} + \frac{\partial \mathcal{H}}{\partial E_k} \delta E_k \right\} dv dt.$$

Recall from Section 5.3 that the stress T_{ij} and electric displacement D_i can be computed using the electric enthalpy density from the identities $T_{ij} = \frac{\partial \mathcal{H}}{\partial S_{ij}}$ and $D_i = -\frac{\partial \mathcal{H}}{\partial E_i}$, and also that the electric field satisfies the electrostatic approximation $E_k = -\frac{\partial \phi}{\partial x_k}$. Furthermore, it follows from properties (V1) and (V3) of Equation 7.2 that we have

$$\delta S_{ij} = \frac{1}{2} \left(\frac{\partial(\delta u_i)}{\partial x_j} + \frac{\partial(\delta u_j)}{\partial x_i} \right),$$
$$\delta E_k = -\frac{\partial}{\partial x_k}(\delta \phi),$$

Also, by the symmetry of the stress tensor T_{ij}, we have

$$\frac{\partial \mathcal{H}}{\partial S_{ij}} \delta S_{ij} = T_{ij} \left\{ \frac{1}{2} \left(\frac{\partial}{\partial x_j} \delta u_i + \frac{\partial}{\partial x_i} \delta u_j \right) \right\}$$
$$= \frac{1}{2}(T_{ij} + T_{ji}) \frac{\partial}{\partial x_j} \delta u_i = T_{ji} \frac{\partial}{\partial x_j} \delta u_i.$$

When these identities are introduced in the variational statement in Equation 7.16 and we enforce property (V4) in Equation 7.2, we find that

$$\delta \int_{t_0}^{t_1} (\mathcal{T} - \mathcal{V}_{\mathcal{H}})dt$$
$$= \int_{t_0}^{t_1} \int_{\Omega} \left\{ -\rho_m \frac{\partial^2 u_i}{\partial t^2} \delta u_i - T_{ji} \frac{\partial}{\partial x_j} \delta u_i - D_i \frac{\partial}{\partial x_i}(\delta \phi) \right\} dv dt$$
$$= \int_{t_0}^{t_1} \int_{\Omega} \left\{ \left(\frac{\partial}{\partial x_j}(T_{ji}) - \rho_m \frac{\partial^2 u_i}{\partial t^2} \right) \delta u_i + \frac{\partial D_i}{\partial x_i} \delta \phi \right\} dv dt$$
$$+ \int_{t_0}^{t_1} \int_{\partial \Omega} \{ -T_{ji} n_j \delta u_i - D_i n_i \delta \phi \} da dt.$$

We obtain the final form of the variational statement for the linearly elastic piezoelectric continuum by integrating by parts to eliminate any derivatives that act on the variations δu_i or $\delta \phi$. We must have

$$\delta \int_{t_0}^{t_1} (\mathcal{T} - \mathcal{V}_{\mathcal{H}})dt + \int_{t_0}^{t_1} \delta W dt$$

$$= \int_{t_0}^{t_1} \int_{\Omega} \left\{ \left(\frac{\partial}{\partial x_j}(T_{ji}) - \rho_m \frac{\partial^2 u_i}{\partial t^2} \right) \delta u_i + \frac{\partial D_i}{\partial x_i} \delta \phi \right\} dv dt$$

$$+ \int_{t_0}^{t_1} \int_{\partial\Omega} \{ (\overline{\tau}_i - T_{ji} n_j) \delta u_i - (\overline{\sigma} + D_i n_i) \delta \phi \} da dt$$

for all admissible variations δu_i and $\delta \phi$ that are consistent with the constraints. We conclude that the mechanical and electrical field variables must satisfy

$$\rho_m \frac{\partial^2 u_i}{\partial t^2} = \frac{\partial T_{ji}}{\partial x_j}, \tag{7.17}$$

$$\frac{\partial D_i}{\partial x_i} = 0, \tag{7.18}$$

for $(t, \boldsymbol{x}) \in \mathbb{R}^+ \times \Omega$, subject to the variational boundary conditions

$$(\overline{\tau}_i - T_{ji} n_j) \delta u_i \Big|_{\partial\Omega} = 0, \tag{7.19}$$

$$(\overline{\sigma} + D_i n_i) \delta \phi |_{\partial\Omega} = 0. \tag{7.20}$$

But Equation 7.17 is the mechanical equilibrium equation and Equation 7.18 is Gauss's Law. These are the two required equations that characterize the initial-boundary value problem for linear piezoelectricity summarized in Equations 5.6 from Section 5.2.2. Equations 7.19 and 7.20 are the associated boundary conditions for the applied stress $\overline{\tau}$ and enforced surface charge distribution $\overline{\sigma}$.

7.3.1 Uniaxial Rod

We consider again the uniaxial rod studied in Examples 6.2 and 7.2.1, but now employ variational methods to derive the equations of motion for the piezoelectric system. If we approximate the electric enthalpy density in this problem as

$$\mathcal{H} \approx \frac{1}{2} C_{33}^E S_{33}^2 - e_{33} S_{33} E_3 - \frac{1}{2} \epsilon_{33}^S E_3^2,$$

the electric enthalpy is obtained by integrating over the volume to yield

$$\mathcal{V}_{\mathcal{H}} = \frac{1}{2} \int_0^L A C_{33}^E \left(\frac{\partial u_3}{\partial x_3} \right)^2 dx_3 + \int_0^L \frac{A e_{33} \chi_{[0,L]}(x_3)}{L} \frac{\partial u_3}{\partial x_3} dx_3 V - \frac{1}{2} \int_0^L A \epsilon_{33}^S \frac{\chi_{[0,L]}}{L^2} dx_3 V^2.$$

In our problem the voltage V is a prescribed input and the unknown is the displacement u_3. By noting that $\delta V = 0$, the Hamilton's principle for linear piezoelectricity requires that

$$0 = \delta \int_{t_0}^{t_1} (\mathcal{T} - \mathcal{V}_{\mathcal{H}}) dt$$

$$= \delta \int_{t_0}^{t_1} (\mathcal{T} - \mathcal{V})dt - \delta \int_{t_0}^{t_1} \int_0^L \frac{Ae_{33}\chi_{[0,L]}}{L} \frac{\partial u_3}{\partial x_3} dx_3 V$$

for all admissible variations δu_3 where $\mathcal{V}$ is the strain energy calculated in Section 7.2.1. When we perform the variation and subsequently integrate by parts to eliminate any derivatives on δu_3, we find the following variational equations:

$$\int_{t_0}^{t_1} \int_0^L \left\{ -\rho_m A \frac{\partial^2 u_3}{\partial t^2} + \frac{\partial}{\partial x_3} \left(AC_{33}^E \frac{\partial u_3}{\partial x_3} \right) + \frac{\partial}{\partial x_3} \left(\frac{Ae_{33}\chi_{[0,L]}}{L} \right) V \right\} \delta u_3 dx_3 dt$$
$$- \int_{t_0}^{t_1} A \left(C_{33}^E \frac{\partial u_3}{\partial x_3} + \frac{e_{33}\chi_{[0,L]}V}{L} \right) \delta u_3 \bigg|_0^L dt = 0$$

We conclude that the displacement $u_3(t, x_3)$ must satisfy the governing equation

$$\rho_m A \frac{\partial^2 u_3}{\partial t^2} = \frac{\partial}{\partial x_3} \left(AC_{33}^E \frac{\partial u_3}{\partial x_3} \right) + \frac{\partial}{\partial x_3} \left(\frac{Ae_{33}\chi_{[0,L]}}{L} \right) V \tag{7.21}$$

for all $(t, x_3) \in \mathbb{R} \times [0, L]$, subject to the variational boundary condition

$$A \left(C_{33}^E \frac{\partial u_3}{\partial x_3} + \frac{e_{33}\chi_{[0,L]}V}{L} \right) \delta u_3 \bigg|_0^L = 0,$$

and subject to the initial conditions

$$u_3(0, x_3) = \overline{u}_3(x_3),$$
$$\frac{\partial u_3}{\partial t} = \overline{v}_3(x_3), x_3 \in [0, L]$$

7.3.2 Bernoulli–Euler Beam

We can use Hamilton's principle for linear piezoelectricity to build on the analysis carried out for linearly elastic beams in Section 7.2.2 and obtain by variational methods the equations of motion of the piezoelectrically actuated beam studied in Section 6.3. For the Bernoulli–Euler beam we approximate the electric enthalpy density as

$$\mathcal{H} \approx \frac{1}{2} C_{11}^E S_{11}^2 - e_{31} S_{11} E_3 - \frac{1}{2} \epsilon_{33}^S E_3^2, \tag{7.22}$$

and we recall that the approximation of the bending strain is given by

$$S_{11} = -x_3 \frac{\partial^2 u_3}{\partial x_1^2}.$$

We further approximate the electric field over the that portion of the beam that supports the piezoelectric surface patches by a piecewise constant function.

$$E_3(x_1,x_3) \approx \begin{cases} -\dfrac{V}{t_p} & (x_1,x_3) \in [a,b] \times [t_s, t_s + t_p] \\ \dfrac{V}{t_p} & (x_1,x_3) \in [a,b] \times [-(t_s+t_p), -t_s] \end{cases}$$

The electric enthalpy is obtained by integrating the electric enthalpy density in Equation 7.22, so that we have

$$\mathcal{V}_H = \frac{1}{2}\int_0^L C_{11}^E I \left(\frac{\partial^2 u_3}{\partial x_1^2}\right)^2 dx_1 + \int_0^L \iint e_{31} x_3 \frac{\partial^2 u_3}{\partial x_1^2} E_3 dx_2 dx_3 dx_1 \\ - \frac{1}{2}\int_0^L \iint \epsilon_{33}^S E_3^2 dx_2 dx_3 dx_1.$$

We can make use of the variational analysis carried out for the linearly elastic beam in Section 6.3 in our study of the piezoelectrically actuated composite beam. By noting that the voltage V in this problem is a prescribed input, and therefore that $\delta V = 0$, it follows that the variational statement for the piezoelectrically actuated beam can be written in the form

$$\delta \int_{t_0}^{t_1} (\mathcal{T} - \mathcal{V}_H) dt \\ = \delta \int_{t_0}^{t_1} (\mathcal{T} - \mathcal{V}) dt + \delta \int_{t_0}^{t_1} \int_0^L \frac{K e_{31} \chi_{[a,b]}}{t_p} \frac{\partial^2 u_3}{\partial x_1^2} dx_1 V$$

where $\mathcal{V}$ is the potential energy that is defined in Section 6.3 for the linearly elastic beam

$$\mathcal{V} := \frac{1}{2}\int_0^L C_{11}^E I \left(\frac{\partial^2 u_3}{\partial x_1^2}\right)^2 dx_1,$$

and $K = K_T - K_B$ is the area moment defined in Section 6.3. The variational statement for the piezoelectrically actuated beam is derived just as that for the linearly elastic beam in Section 7.2.2, but includes additional terms associated with the piezoelectric materials. When we perform the variation $\delta(\dot{})$ and integrate by parts to eliminate derivatives on the variations, we find the following:

$$\int_{t_0}^{t_1} \int_0^L \left\{ -\rho_m A \frac{\partial^2 u_3}{\partial t^2} - \frac{\partial^2}{\partial x_1^2}\left(C_{11}^E I \frac{\partial^2 u_3}{\partial x_1^2} \right) + \frac{\partial^2}{\partial x_1^2}\left(\frac{K e_{31} \chi_{[a,b]}}{t_p} \right) V \right\} \delta u_3 dx_1 dt \\ + \int_{t_0}^{t_1} \left\{ -C_{11}^E I \frac{\partial^2 u_3}{\partial x_1^2} + \frac{K e_{31} \chi_{[a,b]} V}{t_p} \right\} \delta \left(\frac{\partial u_3}{\partial x_1} \right) \Bigg|_0^L dt$$

$$+\int_{t_0}^{t_1} \left\{ \frac{\partial}{\partial x_1} \left(C_{11}^E I \frac{\partial^2 u_3}{\partial x_1^2} \right) - \frac{\partial}{\partial x_1} \left(\frac{K e_{31} \chi_{[a,b]} V}{t_p} \right) \right\} \delta u_3 \big|_0^L \, dt$$

Since this equation must hold for all admissible variations δu, we find that the displacement $u_3(t, x_1)$ must satisfy the equation

$$\rho_m A \frac{\partial^2 u_3}{\partial t^2} = -\frac{\partial^2}{\partial x_1^2} \left(C_{11}^E I \frac{\partial^2 u_3}{\partial x_1^2} \right) + \frac{\partial^2}{\partial x_1^2} \left(\frac{K e_{31} \chi_{[a,b]}}{t_p} \right) V \tag{7.23}$$

for all $(t, x_1) \in \mathbb{R}^+ \times [0, L]$, subject to the variational boundary conditions

$$\left\{ -C_{11}^E I \frac{\partial^2 u_3}{\partial x_1^2} + \frac{K e_{31} \chi_{[a,b]} V}{t_p} \right\} \delta \left(\frac{\partial u_3}{\partial x_1} \right) \Bigg|_0^L = 0$$

$$\left\{ \frac{\partial}{\partial x_1} \left(C_{11}^E I \frac{\partial^2 u_3}{\partial x_1^2} \right) - \frac{\partial}{\partial x_1} \left(\frac{K e_{31} \chi_{[a,b]} V}{t_p} \right) \right\} \delta u_3 \big|_0^L = 0,$$

for $t \in \mathbb{R}^+$, and subject to the initial conditions

$$u_3(0, x_1) = \overline{u}_3(x_1)$$

$$\frac{\partial u_3}{\partial t}(0, x_1) = \overline{v}_3(x_1)$$

for $x_1 \in [0, L]$ when $t = 0$.

7.4 Bernoulli–Euler Beam with a Shunt Circuit

The governing equations for the piezoelectrically actuated beam presented in Sections 6.3 and 7.3.2 have been derived under the assumption that we have prescribed the values of the potential ϕ on the surfaces of the electrodes in Figure 7.1. In many problems, for example in the host of applications summarized in [14] in energy harvesting, the potential ϕ is not prescribed as a boundary condition but rather represents another unknown. Figure 7.1 is a typical configuration in which a piezoelectric structure is connected to a passive shunt circuit. The actual form of the shunt circuit is not a primary concern in this problem: any shunt circuit will suffice for purposes of illustration.

Recall that Hamilton's extended principle for linearly piezoelectric materials in Equation (7.13) holds that of all the admissible trajectories of the piezoelectric system, the true motion satisfies the equation

$$\delta \int_{t_0}^{t_1} (\mathcal{T} - \mathcal{V}_{\mathcal{H}}) dt + \int_{t_0}^{t_1} \delta W dt = 0. \tag{7.24}$$

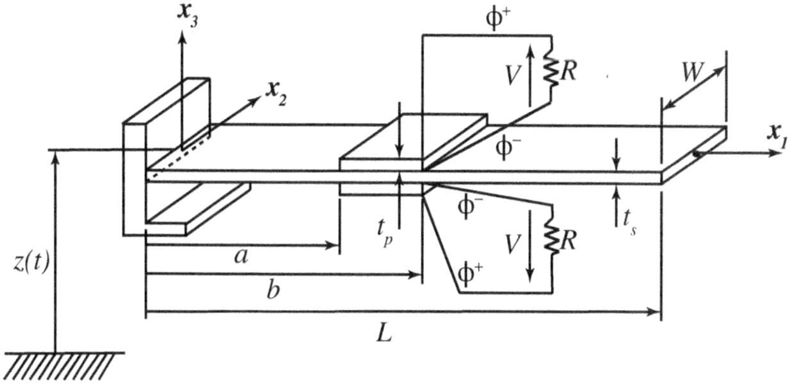

Figure 7.1 Piezoelectric composite beam connected to a passive resistive shunt circuit

As in Section 7.3.2, the electric enthalpy density in Equation 7.22 can be integrated over the beam to obtain the electric enthalpy

$$\mathcal{V}_H = \frac{1}{2}\int_0^L C_{11}^E I\left(\frac{\partial^2 u_3}{\partial x_1^2}\right)^2 dx_1 - \int_0^L \frac{Ke_{31}\chi_{[a,b]}}{t_p} V\frac{\partial^2 u_3}{\partial x_1^2}dx_1$$
$$-\frac{1}{2}\int_0^L A_p\epsilon_{33}^S\chi_{[a,b]}\left(\frac{V}{t_p}\right)^2 dx_1.$$

In contrast to the case considered in Section 7.3.2, however, in this problem both u_3 and V are unknown. The collection of admissible trajectories of the piezoelectric system consists of functions that have the forms $u_3 + \delta u_3$ and $V + \delta V$ where the variations δu_3 and δV are independent. Application of the variational operator $\delta(\cdot)$ to the piezoelectric action integral yields

$$\delta\int_{t_0}^{t_1}(\mathcal{T} - \mathcal{V}_H)dt = \int_{t_0}^{t_1}\int_0^L \left\{\rho_m A\frac{\partial u}{\partial t}\delta\left(\frac{\partial u}{\partial t}\right) - C_{11}^E I\frac{\partial^2 u_3}{\partial x_1^2}\delta\left(\frac{\partial^2 u_3}{\partial x_1^2}\right)\right.$$
$$\left.+\frac{Ke_{31}\chi_{[a,b]}}{t_p}V\delta\left(\frac{\partial^2 u_3}{\partial x_1^2}\right) + \frac{Ke_{31}\chi_{[a,b]}}{t_p}\frac{\partial^2 u_3}{\partial x_1^2}\delta V + \frac{A_p\epsilon_{33}^S\chi_{[a,b]}}{t_p^2}\chi_{[a,b]}V\delta V\right\} dx_1 dt = 0$$

The contribution of the virtual work δW to Equation 7.13 is calculated using the identity introduced in Equation 7.16,

$$\int_{t_0}^{t_1}\delta W dt = -\int_{t_0}^{t_1}\int_{\partial\Omega}\overline{\sigma}\delta\phi da dt.$$

We first consider the contribution of just the top piezoelectric layer, which extends from $x_3 = t_s$ to $x_3 = t_s + t_p$. Without loss of generality, let us suppose that we take $\phi^- = 0$. The potential function and its variation over the top piezoelectric layer is

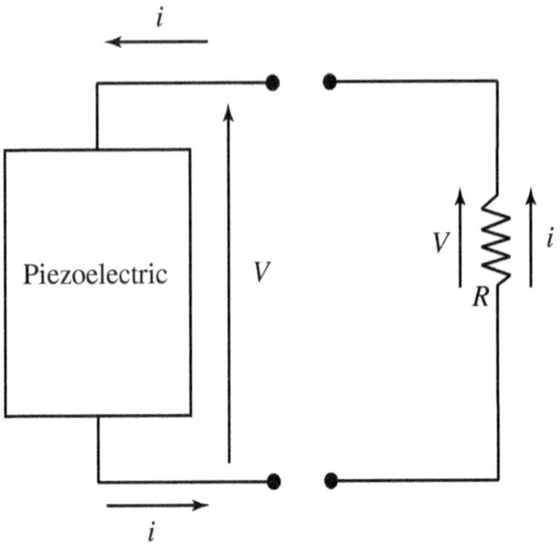

Figure 7.2 Schematic diagram of piezoelectric element with shunt resistor

given by

$$\phi = \frac{x_3 - t_s}{t_p} V \qquad \Longrightarrow \qquad \delta\phi = \frac{x_3 - t_s}{t_p} \delta V.$$

The integration of the virtual work over the upper and lower electrodes that bound the top piezoelectric layer is then

$$\begin{aligned}\int_{t_0}^{t_1} \int_{\partial\Omega} \overline{\sigma} \delta\phi ds dt &= \int_{t_0}^{t_1} \left\{ \int_{\partial\Omega_T} \left(\frac{q}{A_e}\right) \left(\frac{t_s + t_p - t_s}{t_p}\right) \delta V ds \right. \\ &\quad \left. + \int_{\partial\Omega_B} \left(\frac{q}{A_e}\right) \left(\frac{t_s - t_s}{t_p}\right) \delta V ds \right\} dt \\ &= \int_{t_0}^{t_1} q \delta V dt,\end{aligned}$$

where q is the charge that is uniformly distributed over the electrode that has area A_e. It is also possible to deduce the virtual work of the nonconservative electrical loads directly from Figure 7.2.

When the piezoelectric is connected to the shunt resistor, an electrical network is formed. The electric power delivered to the network by the resistor is $P_e = -iV$. The virtual work due to variations δV of the voltage V is equal to $\delta W = -q\delta V$.

A similar calculation over the top and bottom surfaces of the lower piezoelectric layer likewise generates a contribution $\int_{t_0}^{t_1} \int_{\partial\Omega} \overline{\sigma} \delta\phi ds dt = \int_{t_0}^{t_1} q \delta V dt$. We next integrate by parts to eliminate the derivatives on δu_3 that appear in $\delta \int_{t_0}^{t_1} (\mathcal{T} - \mathcal{V}_H) dt$ and combine the result with the virtual work contribution $\int_{t_0}^{t_1} \delta W dt$. We find that we must have

$$\int_{t_0}^{t_1}\int_0^L \left\{ -\rho_m A \frac{\partial^2 u_3}{\partial t^2} - \frac{\partial^2}{\partial x_1^2}\left(C_{11}^E I \frac{\partial^2 u_3}{\partial x_1^2}\right) + \frac{\partial^2}{\partial x_1^2}\left(\frac{Ke_{31}\chi_{[a,b]}}{t_p}\right) V \right\} \delta u_3 dx_1 dt$$
$$+ \int_{t_0}^{t_1} \left\{ \int_0^L \frac{Ke_{31}\chi_{[}a,b]}{t_p}\frac{\partial^2 u_3}{\partial x_1^2} dx_1 + \frac{A_p\epsilon_{33}^S(b-a)}{t_p^2} V - 2q \right\} \delta V dt \tag{7.25}$$
$$+ \int_{t_0}^{t_1} \left\{ -C_{11}^E I \frac{\partial^2 u_3}{\partial x_1^2} + \frac{Ke_{31}\chi_{[a,b]} V}{t_p} \right\} \delta\left(\frac{\partial u_3}{\partial x_1}\right)\Bigg|_0^L dt$$
$$+ \int_{t_0}^{t_1} \left\{ \frac{\partial}{\partial x_1}\left(C_{11}^E I \frac{\partial^2 u_3}{\partial x_1^2}\right) - \frac{\partial}{\partial x_1}\left(\frac{Ke_{31}\chi_{[a,b]} V}{t_p}\right) \right\} \delta u_3\Big|_0^L dt = 0,$$

for all variations δu_3 and δV consistent with the constraints on the system. The displacement $u_3 = u_3(t, x_1)$ and voltage V must satisfy the equations

$$\rho_m A \frac{\partial^2 u_3}{\partial t^2} = -\frac{\partial^2}{\partial x_1^2}\left(C_{11}^E I \frac{\partial^2 u_3}{\partial x_1^2}\right) + \frac{\partial^2}{\partial x_1^2}\left(\frac{Ke_{31}\chi_{[a,b]}}{t_p}\right) V,$$
$$0 = \int_0^L \frac{Ke_{31}\chi_{[a,b]}}{t_p}\frac{\partial^2 u_3}{\partial x_1^2} dx_1 + \frac{A_p(b-a)\epsilon_{33}^S}{t_p^2} V - 2q,$$

for all $(t, x_1) \in \mathbb{R} \times [0, L]$. If, for example the shunt circuit is a simply a resistor, and we define the current as $V = -R\frac{dq}{dt}$, this latter equation can be written as

$$0 = \int_0^L \frac{Ke_{31}\chi_{[a,b]}}{t_p}\frac{\partial^3 u_3}{\partial t \partial x_1^2} dx_1 + \frac{A_p(b-a)\epsilon_{33}^S}{t_p^2}\frac{dV}{dt} + \frac{2}{R} V.$$

These equations are a special case of the general equations of motion for the nonlinear piezoelectric composite beam studied in [42].

We have shown in Section 7.4 that the contribution of resistor to the equations of motion can be derived by calculating the virtual work δW. Alternatively, it is possible to define an electrical dissipation function as

$$\mathcal{D} = \frac{1}{2R_i}\dot{\lambda}_i^2,$$

with the summation over $i = 1, \ldots n_R$. We calculate the contribution to the variational statement as

$$\delta W = -\frac{\partial D}{\partial \dot{\lambda}_k}\partial \lambda_k.$$

Example 7.4.1 Consider again the piezoelectric composite beam with the two shunt resistors depicted in Figure 7.1. The dissipation function $\mathcal{D}$ for the two resistors in this case takes the form

$$\mathcal{D} = \frac{1}{2R}\dot{\lambda}^2 + \frac{1}{2R}\dot{\lambda}^2 = \frac{1}{R}\dot{\lambda}^2,$$

and the virtual work contribution of the two resistors is consequently

$$\delta W = -\frac{2}{R}\dot{\lambda}\delta\lambda.$$

The first two terms on the left in Equation line 7.25 multiply the variation $\delta V = \delta(\dot{\lambda})$, and we want to combine the contribution $\delta W = -\frac{2}{R}\dot{\lambda}\delta\lambda$ due to the resistors with this term. We have the choice of integrating by parts the first two terms in Equation 7.25 so that all three terms multiply $\delta\lambda$, or of integrating the virtual work of the resistors by parts so that all three terms multiply $\delta\dot{\lambda}$. We choose the latter approach. The nonconservative contribution of the resistors to Hamilton's Principle can be calculated by integrating by parts as follows:

$$\begin{aligned}
\int_{t_0}^{t_1} \delta W dt &= -\int_{t_0}^{t_1} \frac{2}{R}\dot{\lambda}\delta\lambda dt \\
&= -\frac{2}{R}\left[\lambda\delta\lambda|_{t_0}^{t_1} - \int_{t_0}^{t_1} \lambda\delta\dot{\lambda}dt\right] = \frac{2}{R}\int_{t_0}^{t_1} \lambda\delta\dot{\lambda}dt \\
&= 2\int_{t_0}^{t_1}\left[\int_{t_0}^{t} \frac{V(\tau)}{R}d\tau + \frac{\lambda_0}{R}\right]\delta V dt \\
&= 2\int_{t_0}^{t_1}\left[\int_{t_0}^{t} (-i(\tau))d\tau + \frac{\lambda_0}{R}\right]\delta V dt \\
&= -2\int_{t_0}^{t_1}\left[\int_{q_0}^{q(t)} dq - \frac{\lambda_0}{R}\right]\delta V dt.
\end{aligned}$$

Finally, when we use the equality $\lambda = -Rq$ in the resistor, we obtain

$$\begin{aligned}
\int_{t_0}^{t_1} \delta W dt &= -\int_{t_0}^{t_1} [2q - 2(q_0 + \frac{\lambda_0}{R})]\delta V dt \\
&= -\int_{t_0}^{t_1} 2q\delta V\, dt.
\end{aligned}$$

This is the same result as calculated from first principles in Equation (7.25).

Example 7.4.2 In this example we derive the equations governing the piezoelectric composite beam with the shunt capacitor that is depicted in Figure 7.3:

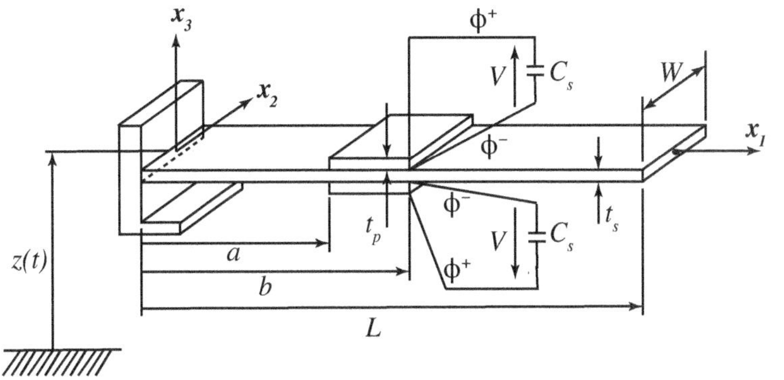

Figure 7.3 Piezoelectric composite beam connected to a passive capacitive shunt circuit

We employ the form of Hamilton's Principle that is applicable to electromechanical systems stated in Theorem 7.3.2. The electromechanical potential $\mathcal{V}_{em}$ is given by the sum

$$\mathcal{V}_{em} = \mathcal{V}_H + \mathcal{V}_m - \mathcal{V}_e,$$

with $\mathcal{V}_m = 0$, $\mathcal{V}_e = 2 \times \left(\frac{1}{2}C_s V^2\right)$, and

$$\mathcal{H} = \frac{1}{2}C_{11}^E S_{11}^2 - e_{31}S_{11}E_3 - \frac{1}{2}\epsilon_{33}^S E_3^2.$$

As in example 7.3.2, the electric enthalpy is given by the quadratic form

$$\mathcal{V}_H = \frac{1}{2}\int_0^L C_{11}^E I\left(\frac{\partial^2 u_3}{\partial x_1^2}\right)^2 dx_1 + \int_0^L \iint e_{31}x_3\frac{\partial^2 u_3}{\partial x_1^2}E_3 dx_2 dx_3 dx_1 - \frac{1}{2}\int_0^L \iint \epsilon_{33}^S E_3^2 dx_2 dx_3 dx_1.$$

In this problem, the virtual work δW of nonconservative electrical elements is zero. The electrical network in Figure 7.4 results when we connect the piezoelectric patch to the shunt capacitor. There are no nonconservative components, so $\delta W = 0$. The variation of the action integral $\mathcal{V}_{em}$ is consequently

(Continued)

Example 7.4.2 (Continued)

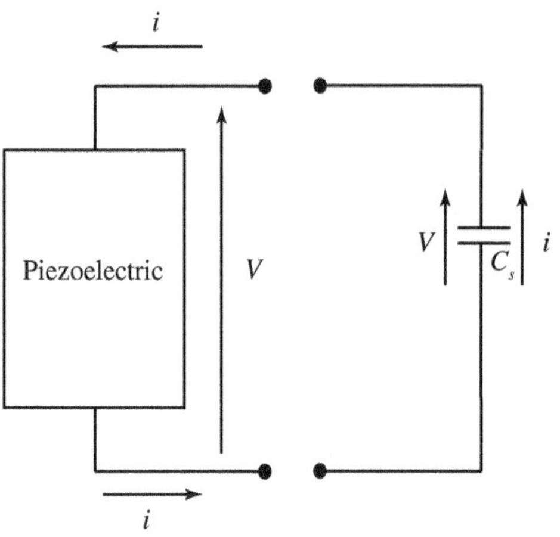

Figure 7.4 Piezoelectric element with capacitive shunt

$$
\begin{aligned}
0 &= \delta \int_{t_0}^{t_1} (T - (\mathcal{V}_{\mathcal{H}} - \mathcal{V}_e))dt,\\
&= \int_{t_0}^{t_1} \int_0^L \left\{ \rho_m A \left(\frac{\partial u}{\partial t} + \dot{z} \right) \delta \left(\frac{\partial u}{\partial t} \right) - C_{11}^E \frac{\partial^2 u_3}{\partial x_1^2} \delta \left(\frac{\partial^2 u_3}{\partial x_1^2} \right) \right.\\
&\quad + \frac{K e_{31} \chi_{[a,b]}}{t_p} V \delta \left(\frac{\partial^2 u_3}{\partial x_1^2} \right) + \frac{K e_{31} \chi_{[a,b]}}{t_p} \frac{\partial^2 u_3}{\partial x_1^2} \delta V\\
&\quad \left. + \frac{A_e \epsilon_{33}^S \chi_{[a,b]}}{t_p^2} \chi_{[a,b]} V \delta V + 2 C_s V \delta V \right\} dx_1 dt.
\end{aligned}
$$

In the usual way, we group the terms into those that multiply the variations δu_3 and δV:

$$
\begin{aligned}
0 &= \int_{t_0}^{t_1} \int_0^L \left\{ -\rho_m A \left(\frac{\partial^2 u_3}{\partial t^2} + \ddot{z} \right) - \frac{\partial^2}{\partial x_1^2} \left(C_{11}^E I \frac{\partial^2 u_3}{\partial x_1^2} \right) \right.\\
&\quad \left. + \frac{\partial^2}{\partial x_1^2} \left(\frac{K e_{31} \chi_{[a,b]}}{t_p} \right) V \right\} \delta u_3 dx_1 dt\\
&\quad + \int_{t_0}^{t_1} \left\{ \int_0^L \frac{K e_{31} \chi_{[a,b]}}{t_p} \frac{\partial^2 u_3}{\partial x_1^2} dx_1 + \frac{A_e \epsilon_{33}^S (b-a)}{t_p^2} V + 2 C_s V \right\} \delta V dt\\
&\quad + \int_{t_0}^{t_1} \left\{ -C_{11}^E I \frac{\partial^2 u_3}{\partial x_1^2} + \frac{K e_{31} \chi_{[a,b]} V}{t_p} \right\} \delta \left(\frac{\partial u_3}{\partial x_1} \right) \Bigg|_0^L dt
\end{aligned}
$$

$$+\int_{t_0}^{t_1}\left\{\frac{\partial}{\partial x_1}\left(C_{11}^E I\frac{\partial^2 u_3}{\partial x_1^2}\right)-\frac{\partial}{\partial x_1}\left(\frac{Ke_{31}\chi_{[a,b]}V}{t_p}\right)\right\}\delta u_3\big|_0^L\,dt.$$

These equations must hold for all the admissible variations δu_3 and δV. The variations δu_3 and δV are arbitrary, so the terms that multiply each must vanish. The strong form of the the governing equations seeks to find the pair (u_3, V) that satisfies the equations

$$\rho_m A\frac{\partial^2 u_3}{\partial t^2}+\frac{\partial^2}{\partial x_1^2}\left(C_{11}^E I\frac{\partial^2 u_3}{\partial x_1^2}\right)-\frac{\partial^2}{\partial x_1^2}\left(\frac{Ke_{31}\chi_{[a,b]}}{t_p}\right)V=-\rho_m A\ddot{z},$$

$$\int_0^L\frac{Ke_{31}\chi_{[a,b]}}{t_p}\frac{\partial^2 u_3}{\partial x_1^2}dx_1+(C_p+2C_s)V=0,$$

with $C_p := \frac{A_e\epsilon_{33}^S(b-a)}{t_p^2}$, subject to the variational boundary conditions

$$0=\left\{-C_{11}^E I\frac{\partial^2 u_3}{\partial x_1^2}+\frac{Ke_{31}\chi_{[a,b]}V}{t_p}\right\}\delta\left(\frac{\partial u_3}{\partial x_1}\right)\Bigg|_0^L,$$

$$0=\left\{\frac{\partial}{\partial x_1}\left(C_{11}^E I\frac{\partial^2 u_3}{\partial x_1^2}\right)-\frac{\partial}{\partial x_1}\left(\frac{Ke_{31}\chi_{[a,b]}V}{t_p}\right)\right\}|\delta u_3{}_0^L,$$

for $x_1 \in [0, L]$, and to the initial conditions

$$u_3(0,x_1)=\overline{u}_3(x_1),$$

$$\frac{\partial u_3}{\partial t}(0,x_1)=\overline{v}_3,$$

for all $x_1 \in [0, L]$.

7.5 Relationship to other Variational Principles

It is important to note that variational formulations of electromechanical systems have been studied extensively and documented in numerous sources. Good overviews can be found in the older works [12, 44], while contemporary accounts can be found in [14, 38, 49]. In all of these accounts of the theory, the notation and definitions vary slightly. In this book we define the electromechanical potential as $\mathcal{V}_{em} := \mathcal{V}_\mathcal{H} + \mathcal{V}_m - \mathcal{V}_e$, with $\mathcal{V}_\mathcal{H}$ the electric enthalpy of a linearly piezoelectric body, $\mathcal{V}_m = \frac{1}{2L_k}i_k^2$ the magnetic potential summing over $k = 1, \ldots, n_L$ discrete ideal inductors, and $\mathcal{V}_e := \frac{1}{2}C_jV_j^2$ the electric potential summing over $j = 1, \ldots, n_C$ discrete ideal capacitors. The most general variational statement introduced in Theorem 7.3.2 states that of all of the possible trajectories of the electromechanical system, the actual motion satisfies

$$\delta \int_{t_0}^{t_1} (\mathcal{T} - \mathcal{V}_{em})dt + \int_{t_0}^{t_1} \delta W dt = 0 \tag{7.26}$$

for all admissible variations of the trajectory in the electromechanical configuration space. In the case when only elastic bodies are included in the electromechanical system, Equation (7.26) can be interpreted as a direct generalization of Hamilton's Principle cast in terms of flux linkage variables in Equation 6.35 of [12] or Equation 3.38 of [38]. If the electromechanical system consists only of piezoelectric bodies, Equation (7.26) is a generalization of Equation 4.79 of [38]. The complexities that can arise when applying this variational approach in some types of problems is discussed in [44]. The author summarizes subtle changes to the variational statement to account for various classes of boundary conditions or internal surfaces of material discontinuity.

The approach followed in this text is similar in philosophy to that in [49] or [44] in that the electric enthalpy plays a central role in the statement of the variational principle. This approach has the advantage that its language thereby closely resembles calculus of variations problems as typically encountered in the study of analytical dynamics, continuum mechanics, or elasticity. At least one potential source of confusion that the reader might encounter in comparing the approach described herein is that the authors in [12] or [38] define two complementary variational principles referred to as the charge/displacement and the flux linkage/displacement formulations. The variational principle in this text is essentially equivalent to the formulation in flux linkage/displacement variables in [38] or [12]. In fact, when [38] demonstrates that that a variational principle induces the correct initial-boundary value problem of linear piezoelectricity for a continuum, he uses the flux linkage/displacement formulation. The proof in Section 4.8 of [38] is essentially the same as that in [44] or in Section 5.2.2 of this text.

A brief review of the approach taken in [12] or [38] will help clarify the similarity of the flux linkage/displacement formulation with the variational approach in this book. The charge/displacement formulation of [12] or [38] is not discussed in this text and the interested reader should consult these references for a full account of that methodology. Reference [38] states that Hamilton's Principle for a piezoelectric continuum in the flux linkage/displacement formulation requires that

$$\delta \int_{t_0}^{t_1} (\mathcal{T}^* + W_e^*)dt + \int_{t_0}^{t_1} \delta W dt = 0 \tag{7.27}$$

for all admissible flux linkage variations $\delta\lambda_k$ and displacement variations δu of the actual trajectory in the electromechanical configuration space. In this equation $\mathcal{T}^*$ is referred to as the "kinetic co-energy." Their definition of the kinetic energy $\tilde{\mathcal{T}}$, which is expressed in terms of the momentum p, is related to the kinetic coenergy

$\mathcal{T}^*$ by a Legendre transform. Recall from Section 5.3 that the Legendre transform is a way, made popular in the field of thermodynamics, to switch the independent variable in families of functions. For a single degree of freedom system such as a point mass, [38] and [12] define $\mathcal{T}^*(v) = pv - \tilde{\mathcal{T}}(p)$ where v is the velocity and p is the linear momentum of the point mass. Those familiar with Hamiltonian mechanics will recognize that the kinetic energy $\tilde{\mathcal{T}}$ so defined coincides with the kinetic energy that appears in a Hamiltonian formulation of mechanics, in contrast to the kinetic energy that appears in a Lagrangian formulation of mechanics. See [17], Chapter 8 for the theoretical justification. In any event, for our purposes and in this book, it suffices to note that the kinetic coenergy $\mathcal{T}^*$ of [38] and [12] coincides with the definition of the Lagrangian kinetic energy $\mathcal{T}$ in this book, so that $\mathcal{T} = \mathcal{T}^*$. Also, as noted on page 120 of [38], the electric coenergy W_e^* for a piezoelectric body is exactly equal to the negative of the electric enthalpy $W_e^* = -\int_\Omega \mathcal{H} dv$. It is then evident that the version of Hamilton's Principle in [38], for example, coincides with that stated in Theorem 7.3.1. Similarly, a version of Hamilton's Principle for electromechanical systems assembled from discrete ideal electrical components, with mechanical properties represented by a collection of discrete mechanical degrees of freedom, but which does not contain distributed piezoelectric continua, is given on page 72 of [38]. This variational statement requires that

$$\delta \int_{t_0}^{t_1} (\mathcal{T}^* + W_e^* - \mathcal{V} - W_m) dt + \int_{t_0}^{t_1} \delta W \, dt = 0 \tag{7.28}$$

for all admissible flux linkage variations $\delta\lambda_k$ and virtual variations δx. In this equation x contains the coordinates of the finite, discrete collection of degrees of freedom of the mechanical system. Also, the electrical coenergy $W_e^* = \frac{1}{2} C_k \dot{\lambda}_k^2$ is given by the summation over $k = 1, \ldots, n_C$ capacitors, and the magnetic energy $W_m = \frac{1}{2L_j} \lambda_j^2$ is summed over $j = 1, \ldots, n_L$ discrete ideal inductors. Note that Theorem 7.3.2 effectively combines the cases treated separately in Equations (7.27) and (7.28).

While the symbols, nomenclature, and naming conventions differ substantially in some of these formulations, they all generate the same equations ultimately. In some cases, it can take considerable work to show that an unfamiliar form of the variational statement is equivalent to another.

Example 7.5.1 It is frequently the case that authors in research papers introduce a variational principle that does not appear to match the forms in works such as [12, 38, 44, 49], or [14]. For example, in [42], the authors employ a variational statement that is cast in terms of a Lagrangian $\mathfrak{L} := \mathcal{T} - \mathfrak{U}$ where

(Continued)

Example 7.5.1 (Continued)

the "total conservative potential $\mathfrak{U}$" is defined as

$$\mathfrak{U} := \frac{1}{2}\int_{\Omega} \mathcal{U}_0 dv + \frac{1}{2}\int_{\Omega_p} (T_{ij}S_{ij} - D_i E_i) dv. \tag{7.29}$$

In this equation the variable $\mathcal{U}_0$ is the strain energy density introduced in Chapter 3, Ω is the volume occupied by the entire specimen, and Ω_p is the subdomain contained in Ω that is occupied by piezoelectric material. If the domain Ω contains only linearly elastic or linearly piezoelectric materials, this variational principle can readily be shown to be equivalent to those that are expressed in terms of the electric enthalpy. Recall that the electric enthalpy density for a linearly piezoelectric body is written as

$$\mathcal{H} := \frac{1}{2}C^E_{ijkl}S_{ij}S_{kl} - e_{mij}E_m S_{ij} - \frac{1}{2}\epsilon^S_{ij}E_i E_j.$$

It has been noted in Chapter 5 that the electric enthalpy density $\mathcal{H}$ for a linearly piezoelectric material reduces to the strain energy $\mathcal{U}_0 = \frac{1}{2}C^E_{ijkl}S_{ij}S_{kl}$ if we set the terms that involve the electric field to zero in the above equation. In this case the first integral can be subsumed into the second integral in Equation (7.29) so that

$$\mathfrak{U} = \frac{1}{2}\int_{\Omega} (T_{ij}S_{ij} - D_i E_i) dv.$$

We substitute the constitutive laws for a linearly piezoelectric material and find that

$$\begin{aligned}
\mathfrak{U} &= \int_{\Omega} \left(\frac{1}{2}(C^E_{ijkl}S_{kl} - e_{mij}E_m)S_{ij} - \frac{1}{2}(e_{ikl}S_{kl} + \epsilon^S_{ij}E_j)E_i \right) dv, \\
&= \int_{\Omega} \left(\frac{1}{2}C^E_{ijkl}S_{ij}S_{kl} - e_{mij}E_m S_{ij} - \frac{1}{2}\epsilon^S_{ij}E_i E_j \right) dv, \\
&= \int_{\Omega} \mathcal{H} dv.
\end{aligned}$$

We conclude that, for a domain constructed from linearly elastic and linearly piezoelectric material, the functional $\mathfrak{U}$ in Equation (7.29) is the electric enthalpy.

7.6 Lagrangian Densities

In this section we review the construction of variational calculus problems in which the action integral $\mathcal{A} := \int_{t_0}^{t_1} \mathcal{L} dt$ is obtained by integration in time of a

Lagrangian $\mathcal{L} := \mathcal{T} - \mathcal{V}_{em}$, which in turn is defined via a Lagrangian density $\mathfrak{L}$,

$$\mathcal{L} := \mathcal{T} - \mathcal{V}_{em} = \int_{\Omega} \mathfrak{L} dv,$$

that is integrated over some spatial domain $\Omega \subset \mathbb{R}^d$. In this text we will only consider the case where $d = 1$, which suffices to treat the composite piezoelectric rod and beam. See [17] Chapter 12, Section 2, for a full exposition of the general case. In many of our cases, we assume that the Lagrangian density $\mathfrak{L}$ over the domain $\Omega = [0, L] \subset \mathbb{R}^1$ has the functional dependence

$$\mathfrak{L} = \mathfrak{L}\left(\frac{\partial u}{\partial t}, u, \frac{\partial u}{\partial x}, \frac{\partial^2 u}{\partial x^2}, V\right) \tag{7.30}$$

where u is the spatial displacement, V is the voltage, and (t, x) are the independent time and space variables. A slightly different functional dependence in $\mathfrak{L}$ is considered in Example 7.6.3. Hamilton's Principle for electromechanical systems which have the functional dependence specified in Equation (7.30) seeks a solution pair (u, V) that satisfies

$$0 = \delta \int_{t_0}^{t_1} \int_0^L \mathfrak{L}\left(\frac{\partial u}{\partial t}, u, \frac{\partial u}{\partial x}, \frac{\partial^2 u}{\partial x^2}, V\right) dxdt, \tag{7.31}$$

for all admissible variations of the trajectory in electromechanical configuration space. We simplify the notation in the remainder of this section by defining

$$\begin{aligned}
\dot{(\cdot)} &:= \frac{\partial}{\partial t}(\cdot), \\
(\cdot)' &:= \frac{\partial}{\partial x}(\cdot), \\
(\cdot)'' &:= \frac{\partial^2}{\partial x^2}(\cdot).
\end{aligned}$$

The stationarity condition in Equation (7.31) implies that

$$\begin{aligned}
0 &= \int_{t_0}^{t_1} \int_0^L \left(\frac{\partial \mathfrak{L}}{\partial \dot{u}} \delta \dot{u} + \frac{\partial \mathfrak{L}}{\partial u} \delta u + \frac{\partial \mathfrak{L}}{\partial u'} \delta u' + \frac{\partial \mathfrak{L}}{\partial u''} \delta u'' + \frac{\partial \mathfrak{L}}{\partial V} \delta V \right) dxdt, \\
&= \int_{t_0}^{t_1} \int_0^L \left\{ -\frac{\partial}{\partial t}\left(\frac{\partial \mathfrak{L}}{\partial \dot{u}}\right) + \frac{\partial \mathfrak{L}}{\partial u} - \frac{\partial}{\partial x}\left(\frac{\partial \mathfrak{L}}{\partial u'}\right) + \frac{\partial^2}{\partial x^2}\left(\frac{\partial \mathfrak{L}}{\partial u''}\right) \right\} \delta u dxdt \\
&+ \int_{t_0}^{t_1} \int_0^L \frac{\partial \mathfrak{L}}{\partial V} \delta V dxdt + \int_{t_0}^{t_1} \left\{ \left(\frac{\partial \mathfrak{L}}{\partial u'} - \frac{\partial}{\partial x}\left(\frac{\partial \mathfrak{L}}{\partial u''}\right) \right) \delta u + \frac{\partial \mathfrak{L}}{\partial u''} \delta u' \right\} |_0^L dt,
\end{aligned}$$

must hold for all admissible variations δu, $\delta u'$, and δV, when we integrate by parts in the usual fashion for variational problems. We conclude that the governing equations are

$$0 = -\frac{\partial}{\partial t}\left(\frac{\partial \mathfrak{L}}{\partial \dot{u}}\right) + \frac{\partial \mathfrak{L}}{\partial u} - \frac{\partial}{\partial x}\left(\frac{\partial \mathfrak{L}}{\partial u'}\right) + \frac{\partial^2}{\partial x^2}\left(\frac{\partial \mathfrak{L}}{\partial u''}\right), \tag{7.32}$$

$$0 = \frac{\partial \mathfrak{L}}{\partial V}, \tag{7.33}$$

subject to the variational boundary conditions

$$0 = \left(\frac{\partial \mathfrak{L}}{\partial u'} - \frac{\partial}{\partial x} \left(\frac{\partial \mathfrak{L}}{\partial u''} \right) \right) \delta u \,\Big|_0^L , \tag{7.34}$$

$$0 = \frac{\partial \mathfrak{L}}{\partial u''} \delta u' \,\Big|_0^L , \tag{7.35}$$

and to the appropriate initial conditions when $t = 0$.

As a first example application of these principles, we re-derive the equations of motion of the axial piezoelectric actuator using a Lagrangian density that is integrated over the domain $\Omega = [0, L]$. In this formulation we approximate the scalar potential ϕ as a piecewise linear function that satisfies the potential boundary conditions.

Example 7.6.1 For the piezoelectric composite rod, the Lagrangian density is given by

$$\mathfrak{L} = \frac{1}{2} \rho_m A \dot{u}^2 - \left(\frac{1}{2} A C_{33}^E (u')^2 + \frac{A e_{33} \mathcal{X}_{[a,b]}}{b-a} u' V - \frac{1}{2} \frac{A \varepsilon_{33}^S \mathcal{X}_{[a,b]}}{b-a} V^2 \right).$$

Equation 7.32 yields

$$0 = -\rho_m A \frac{\partial^2 u_3}{\partial t^2} + \frac{\partial}{\partial x_3} \left(A C_{33}^E \frac{\partial u_3}{\partial x_3} \right) + \frac{\partial}{\partial x_3} \left(\frac{A e_{33} \mathcal{X}_{[a,b]}}{b-a} \right) V, \tag{7.36}$$

Equation 7.33 generates the equation

$$0 = \frac{A e_{33} \mathcal{X}_{[a,b]}}{b-a} \frac{\partial u_3}{\partial x_3} - \frac{A \epsilon_{33}^S}{(b-a)^2} V, \tag{7.37}$$

and the variational boundary condition 7.34 reduces to

$$0 = \left(A C_{33}^E \frac{\partial u_3}{\partial x_3} + \frac{A e_{33} \mathcal{X}_{[a,b]}}{b-a} V \right) \Big|_0^L . \tag{7.38}$$

The first Equation 7.36 is identical to Equation 7.21, and the variational boundary condition in Equation 7.38 is the same as in Section 7.3.1. Since we have assumed that δV is arbitrary in this problem, in contrast to Section 7.3.1 where $\delta V = 0$, we obtain a new Equation 7.37 in this example. It is straightforward to show that Equation 7.37 can also be obtained by enforcing the prescribed surface charge to be zero via the boundary condition $0 = q = \int \overline{\sigma} da = -\int D_3 n_3 da$, and then subsequently substituting $D_3 = e_{33} S_{33} + \epsilon_{33}^S E_3$.

In the next example we similarly derive the equations of motion of the composite piezoelectric beam when we make the assumption that the scalar potential ϕ is approximated by a piecewise linear function that satisfies the potential boundary conditions.

Example 7.6.2 For the Bernoulli–Euler beam, when we approximate the scalar potential ϕ by a piecewise linear function that satisfies the potential boundary conditions, the Lagrangian density $\mathfrak{L}$ is given by

$$\mathfrak{L} = \frac{1}{2}\rho_m A(\dot{u}_3)^2 - \left\{ \frac{1}{2} C_{11}^E I (u'')^2 - \frac{\kappa e_{31} \chi_{[a,b]}}{t_p} V u'' - \frac{1}{2} \frac{A_p \epsilon_{33}^S \chi_{[a,b]}}{t_p^2} V^2 \right\},$$

We apply the general equations of motion in Equations 7.32 through 7.35 to determine the equations of motion for this particular case. We have

$$\begin{aligned} 0 = & \int_{t_0}^{t_1} \int_0^L \left\{ -\rho_m A \frac{\partial^2 u_3}{\partial t^2} + \frac{\partial^2}{\partial x_1^2} (-C_{11}^E I \frac{\partial^2 u_3}{\partial x_1^2} + \frac{\kappa e_{31} \chi_{[a,b]}}{t_p} V) \right\} \delta u_3 dx_1 dt \\ & + \int_{t_0}^{t_1} \left\{ \int_0^L \frac{\kappa e_{31} \chi_{[a,b]}}{t_p} \frac{\partial^2 u_3}{\partial x_1^2} dx + \frac{A_p \epsilon_{33}^S (b-a)}{t_p^2} V \right\} \delta V dt \\ & + \int_{t_0}^{t_1} \left\{ \frac{\partial}{\partial x_1} \left(C_{11}^E I \frac{\partial^2 u_3}{\partial x_1^2} \right) - \frac{\partial}{\partial x_1} \left(\frac{\kappa e_{31} \chi_{[a,b]}}{t_p} \right) V \right\} \delta u_3 |_0^L dt \\ & + \int_{t_0}^{t_1} \left\{ -C_{11}^E I \frac{\partial^2 u_3}{\partial x_1^2} + \frac{\kappa e_{31} \chi_{[a,b]}}{t_p} V \right\} \delta \left(\frac{\partial u_3}{\partial x_1} \right) |_0^L dt \end{aligned}$$

A careful study of these equations shows that they are identical to the equations derived in Example 7.4.2 when we set $C_s = 0$.

As a last example of the use of Lagrangian densities, we derive the equations governing the motion of the axial piezoelectric composite actuator, but in this case do not approximate the potential ϕ as a piecewise linear function.

Example 7.6.3 The Lagrangian density for the piezoelectric composite rod takes the form

$$\mathfrak{L} = \frac{1}{2}\rho_m A(\dot{u}_3)^2 - \left\{ \frac{1}{2} C_{33}^E I (u')^2 + A e_{33} \chi_{[a,b]} u' \phi' - \frac{1}{2} A \epsilon_{33}^S (\phi')^2 \right\},$$

This Lagrangian density has a slightly different functional dependence than the last two examples, $\mathfrak{L} = \mathfrak{L}(\dot{u}, u, u', u'', \phi, \phi')$. The stationarity condition for the action integral now takes the form

$$0 = \int_{t_0}^{t_1} \int_0^L \left(\frac{\partial \mathfrak{L}}{\partial \dot{u}} \delta \dot{u} + \frac{\partial \mathfrak{L}}{\partial u} \delta u + \frac{\partial \mathfrak{L}}{\partial u'} \delta u' \right.$$

(Continued)

Example 7.6.3 (Continued)

$$
\begin{aligned}
&+\frac{\partial \mathfrak{L}}{\partial u''}\delta u'' + \frac{\partial \mathfrak{L}}{\partial \phi}\delta\phi + \frac{\partial \mathfrak{L}}{\partial \phi'}\delta\phi' \Bigg) dxdt, \\
&= \int_{t_0}^{t_1}\int_0^L \left\{ -\frac{\partial}{\partial t}\left(\frac{\partial \mathfrak{L}}{\partial \dot{u}}\right) + \frac{\partial \mathfrak{L}}{\partial u} - \frac{\partial}{\partial x}\left(\frac{\partial \mathfrak{L}}{\partial u'}\right) + \frac{\partial^2}{\partial x^2}\left(\frac{\partial \mathfrak{L}}{\partial u''}\right) \right\} \delta u dxdt \\
&+ \int_{t_0}^{t_1}\int_0^L (\frac{\partial \mathfrak{L}}{\partial \phi} - \frac{\partial}{\partial x}\left(\frac{\partial \mathfrak{L}}{\partial \phi'}\right))\delta\phi dxdt \\
&+ \int_{t_0}^{t_1} \left\{ \left(\frac{\partial \mathfrak{L}}{\partial u'} - \frac{\partial}{\partial x}\left(\frac{\partial \mathfrak{L}}{\partial u''}\right)\right) \delta u + \frac{\partial \mathfrak{L}}{\partial u''}\delta u' + \frac{\partial \mathfrak{L}}{\partial \phi'}\delta\phi|_0^L \right\} |_0^L dt,
\end{aligned}
$$

These equations can be applied directly to the problem at hand. A solution pair (u_3, ϕ) must satisfy

$$
\begin{aligned}
0 &= -\rho_m A\frac{\partial u_3}{\partial t^2} - \frac{\partial}{\partial x_3}\left(-AC_{33}\frac{\partial u_3}{\partial x_3} - Ae_{33}\chi_{[a,b]}\frac{\partial \phi}{\partial x_3}\right), \\
0 &= -\frac{\partial}{\partial x_3}\left(-Ae_{33}\chi_{[a,b]}\frac{\partial u_3}{\partial x_3} + A\epsilon_{33}^S\chi_{[a,b]}\frac{\partial \phi}{\partial x_3}\right),
\end{aligned}
$$

subject to the boundary conditions

$$
\begin{aligned}
0 &= \left(-AC_{33}^E\frac{\partial u_3}{\partial x_3} - Ae_{33}\chi\frac{\partial \phi}{\partial x_3}\right)\delta u_3|_0^L, \\
0 &= -\frac{\partial}{\partial x_3}\left(-Ae_{33}\chi_{[a,b]}\frac{\partial u_3}{\partial x_3} + A\epsilon_{33}^S\chi_{[a,b]}\frac{\partial \phi}{\partial x_3}\right)\delta\phi|_0^L,
\end{aligned}
$$

and to the appropriate initial conditions.

Example 7.6.4 In this example we derive the equations of motion for the stack actuator shown schematically in Figure 7.5 that is driven by a prescribed voltage $V(t)$.

Note that the n layers of the stack actuator are poled in the direction marked with an arrow, and the direction of poling alternates from one layer to the next. The width of one layer $t_p = \frac{L}{n}$. We use a piecewise linear approximation of the electrical potential $\phi = \phi(x_3)$, and as a consequence the electric field in the x_3 direction has the form

$$
E_3 := E_3(x_3) = (-1)^i\frac{V}{L/n} \qquad x_3 \in \left[\frac{(i-1)L}{n}, \frac{iL}{n}\right]
$$

for $i = 1, \ldots, n$. We write the constitutive law for an odd numbered layer in the form

$$
\begin{Bmatrix} T_{33} \\ D_3 \end{Bmatrix} = \begin{bmatrix} C_{33}^E & -e_{33} \\ e_{33} & \epsilon_{33}^S \end{bmatrix} \begin{Bmatrix} S_{33} \\ E_3 \end{Bmatrix}.
$$

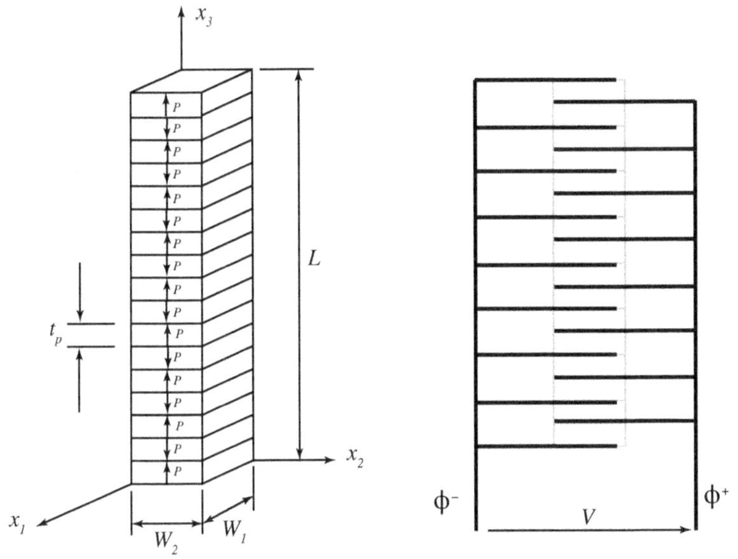

Figure 7.5 Stack actuator, orientation of layers, and electroding pattern

where the tensor components T_{ij}, S_{ij}, D_i, E_i are given with respect to the x_1, x_2, x_3 axes. The constitutive law for an even numbered layer are also written in form

$$\begin{Bmatrix} T'_{33} \\ D'_3 \end{Bmatrix} = \begin{bmatrix} C^E_{33} & -e_{33} \\ e_{33} & \epsilon^S_{33} \end{bmatrix} \begin{Bmatrix} S'_{33} \\ E'_3 \end{Bmatrix}.$$

where the components of the tensors $T'_{ij}, S'_{ij}, E'_i, D'_i$ are defined with respect to their local coordinates x'_1, x'_2, x'_3. The even layer, local coordinates x'_1, x'_2, x'_3 are obtained from the odd layer, global coordinates x_1, x_2, x_3 by rotation about the $x_1 = x'_1$ axis through π. The rotation matrix from odd to even layer bases is given by

$$\begin{Bmatrix} x'_1 \\ x'_2 \\ x'_3 \end{Bmatrix} = \begin{bmatrix} 1 & 0 & 0 \\ 0 & -1 & 0 \\ 0 & 0 & -1 \end{bmatrix} \begin{Bmatrix} x_1 \\ x_2 \\ x_3 \end{Bmatrix}.$$

The orientation of frames of reference for the odd and even layers is shown in Figure 7.6.

(Continued)

Example 7.6.4 (Continued)

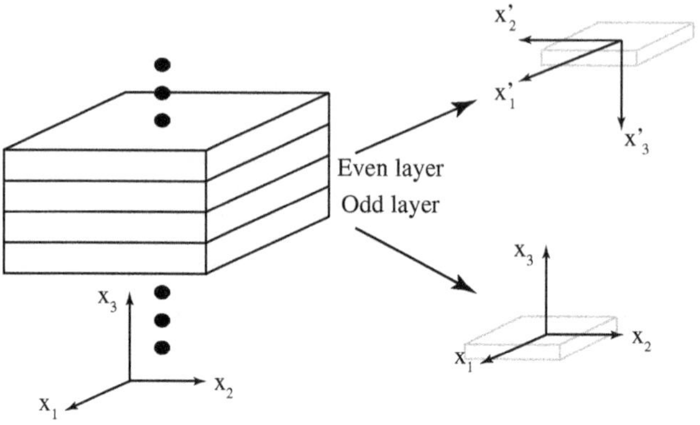

Figure 7.6 Coordinate systems for odd and even layers

It is also possible to choose the rotation matrix above as a rotation of π about the x_2 axis, or a reflection about the $x_1 - x_2$ plane, with no change in the conclusions that follow. Recall that e'_{33} is the compact notation for the tensor component e'_{333}, and that we transform this component according to the transformation law $e_{333} = r_{33}r_{33}r_{33}e'_{333} = (-1)(-1)(-1)e'_{333}$. We find that in an even layer of the stack that $e_{33} = -e'_{33}$. We can argue similarly to show that $C^E_{33} = C^{E,'}_{33}$ and $\epsilon^S_{33} = \epsilon^{S,'}_{33}$. When we transform the even layer constitutive law in terms of components relative to the basis x_1, x_2, x_3, we find that

$$\left\{ \begin{array}{c} T_{33} \\ D_3 \end{array} \right\} = \begin{bmatrix} C^E_{33} & e_{33} \\ -e_{33} & \epsilon^S_{33} \end{bmatrix} \left\{ \begin{array}{c} S_{33} \\ E_3 \end{array} \right\}.$$

In summary, we can write a single expression

$$\left\{ \begin{array}{c} T_{33} \\ D_3 \end{array} \right\} = \begin{bmatrix} C^E_{33} & (-1)^i e_{33} \\ (-1)^{i+1} e_{33} & \epsilon^S_{33} \end{bmatrix} \left\{ \begin{array}{c} S_{33} \\ E_3 \end{array} \right\} \quad \begin{array}{l} x_3 \in \left[\dfrac{(i-1)L}{n}, \dfrac{iL}{n} \right], \\ i = 1, \dots, n, \end{array}$$

that gives the spatially varying constitutive law for all $x_3 \in [0, L]$. The electric enthalpy density is given as a function of x_3 by the expression

$$\begin{aligned} \mathcal{H} &\approx \frac{1}{2} C^E_{33} S^2_{33} - e_{33} S_{33} E_3 - \frac{1}{2} \epsilon^S_{33} E^2_3 \\ &= \frac{1}{2} C^E_{33} S^2_{33} + \left((-1)^i e_{33} \right) S_{33} \left((-1)^i \frac{V}{L/n} \right) - \frac{1}{2} \epsilon^S_{33} \left(\frac{V}{L/n} \right)^2 \end{aligned}$$

for $x_3 \in [\frac{(i-1)L}{n}, \frac{iL}{n}]$ with $i = 1, \ldots, n$. The electric enthalpy is obtained by integrating over the volume to yield

$$\mathcal{V}_{\mathcal{H}} = \frac{1}{2}\int_0^L AC_{33}^E\left(\frac{\partial u_3}{\partial x_3}\right)^2 dx_3 + \int_0^L \frac{Ae_{33}\chi_{[0,L]}(x_3)}{L}\frac{\partial u_3}{\partial x_3}dx_3 nV \tag{7.39}$$

$$-\frac{1}{2}\int_0^L A\epsilon_{33}^S \frac{\chi_{[0,L]}}{L^2}dx_3 n^2 V^2. \tag{7.40}$$

The final equation of motion

$$\rho_m A\frac{\partial^2 u_3}{\partial t^2} = \frac{\partial}{\partial x_3}\left(AC_{33}^E\frac{\partial u_3}{\partial x_3}\right) + \frac{\partial}{\partial x_3}\left(\frac{Ae_{33}\chi_{[0,L]}}{L}\right)nV,$$

is obtained by differentiating the Lagrangian $\mathfrak{L}$ in

$$\mathfrak{L} = \frac{1}{2}\rho_m A\frac{\partial^2 u_3}{\partial t^2} - \left(\frac{1}{2}AC_{33}^E\left(\frac{\partial u_3}{\partial x_3}\right)^2 + \frac{Ae_{33}\chi_{[0,L]}(x_3)}{L}\frac{\partial u_3}{\partial x_3}nV\right.$$
$$\left.-\frac{1}{2}A\epsilon_{33}^S\frac{\chi_{[0,L]}}{L^2}n^2V^2\right).$$

The purpose of alternating thin layers and using the electrode pattern shown above is now evident in Equation 7.40. We see that the stack actuator equation of motion is drive by the input $nV(t)$, while a monolithic, single layer, axial actuator is driven by the input voltage $V(t)$. Since the equations are otherwise identical, and they are linear, the response of the stack actuator will have an amplitude that is a factor n times that of the single layer specimen for the same input voltage $V(t)$, boundary conditions, and initial conditions.

7.7 Problems

Problems 7.7.1 Use Hamilton's Principle to derive the governing equations for linearly elastic rod with a tip mass.

Problems 7.7.2 Use Hamilton's Principle to derive the governing equations for a linearly elastic Bernoulli–Euler beam with a tip mass.

Problems 7.7.3 Use Hamilton's Principle to derive the governing equations for a linearly elastic rod with a moving base.

Problems 7.7.4 Use Hamilton's Principle to derive the governing equations for a linearly elastic Bernoulli–Euler beam with a moving base.

Problems 7.7.5 Use Hamilton's Principle to derive the governing equations for linearly piezoelectric rod with a tip mass depicted in Figure 6.5.1.

Problems 7.7.6 Use Hamilton's Principle to derive the governing equations for the linearly piezoelectric Bernoulli–Euler composite beam with a tip mass depicted in Problem 6.5.3.

Problems 7.7.7 Use Hamilton's Principle to derive the governing equations for a linearly piezoelectric rod with a moving base depicted in Figure 6.5.2.

Problems 7.7.8 Use Hamilton's Principle to derive the governing equations for a linearly piezoelectric Bernoulli–Euler beam with a moving base depicted in Figure 6.5.4.

Problems 7.7.9 Repeat Problem 7.5 using the Lagrangian density and the variational technique summarized in Section 7.6.

Problems 7.7.10 Repeat Problem 7.6 using the Lagrangian density and the variational technique summarized in Section 7.6.

Problems 7.7.11 Repeat Problem 7.7 using the Lagrangian density and the variational technique summarized in Section 7.6.

Problems 7.7.12 Repeat Problem 7.8 using the Lagrangian density and the variational technique summarized in Section 7.6.

8

Approximations

Chapters 6 and 7 have derived partial differential equations that govern the time evolution of a variety of typical or canonical piezoelectric structures. In each case the unknowns evolve in an infinite dimensional state space of functions. In practical applications we must use approximations of these governing equations to calculate estimates of the system response. In this chapter we describe two of the most common approaches for deriving approximate solutions: modal or eigenfunction techniques and finite element methods. The chapter begins with a discussion of the classical, strong, and weak solutions of the governing equations in Section 8.1. We then discuss Galerkin approximations of the weak form of the governing equations in Section 8.3. Section 8.3.1 reviews how the eigenfunctions of the spatial differential operators that appear in the governing equations can be used as the basis functions in a Galerkin approximation, while Section 8.3.2 details the construction of the approximating basis from simple finite element shape functions. To make the discussions concrete, numerous examples throughout the chapter outline the fundamentals of building approximations for the composite piezoelectric axial rod and beam.

8.1 Classical, Strong, and Weak Formulations

Equations 6.11 and 7.21 for an axial rod, Equations 6.20 and 7.23 for a beam, or Equation 6.32 for a plate are iconic examples of piezoelectric composite structures. Each of these partial differential equations can be written in a general form discussed in Appendix S.1.2. In our discussion of the approximations in the next few sections, we will cast formulations in terms of displacement functions u that depend on the time $t \in \mathbb{R}^+$ and spatial coordinates $x \in \Omega \subset \mathbb{R}^d$, with Ω a problem-specific spatial domain. For the axial and bending composites, $d = 1$ and $\Omega = [0, L] \subset \mathbb{R}$. For the square piezoelectric plate, $d = 2$ and $\Omega = [0, L] \times [0, L] \subset \mathbb{R}^2$. See [4] for more complex cases in terms of vector-valued

Vibrations of Linear Piezostructures, First Edition. Andrew J. Kurdila and Pablo A. Tarazaga.

This Work is a co-publication between John Wiley & Sons Ltd and ASME Press.

displacement fields over irregular domains contained in $\mathbb{R}^2$ and $\mathbb{R}^3$. We denote $u(t,x)$ the value of the function u at time $t \in \mathbb{R}^+$ at the location $x \in \Omega$. It is conventional in the study of distributed parameter systems (DPS) to represent the displacement state by $u(t) := u(t,\cdot)$, which is a function of $x \in \Omega$ at each fixed time $t \in \mathbb{R}^+$. The state $u(t)$ is said to be infinite dimensional since at each time $\in \mathbb{R}^+$ it belongs to an infinite dimensional space of functions in the spatial variable $x \in \Omega$. With these notational conventions, we seek a displacement field u that satisfies

$$\mathfrak{M}\,\frac{\partial^2 u}{\partial t^2}(t) + \mathfrak{C}\frac{\partial u}{\partial t}(t) + \mathfrak{K}u(t) = f(t) \tag{8.1}$$

at time $t \in \mathbb{R}^+$ over the domain $\Omega \subset \mathbb{R}^d$, as well as suitable boundary and initial conditions, when $\mathfrak{M}$ is a spatial multiplication operator representing the mass distribution, $\mathfrak{C}$ is a differential operator representing the distributed damping, $\mathfrak{K}$ is a spatial stiffness differential operator representing the distributed stiffness, and $f(t) := f(t,\cdot)$ is the input spatially distributed load function at time $t \in \mathbb{R}^+$. For the piezoelectric composite examples in Equations 6.11 and 6.20, we initially consider an undamped model with $\mathfrak{C} \equiv 0$ and a specific form for the distributed mass operator $\mathfrak{M}$ and loading f. The discussion of various models of damping is given in Section 8.2 and in the Appendix in Section 1.2.

So in our first few models, we seek a displacement field $u(t) := u(t,\cdot)$ that obeys the equation

$$\rho A\frac{\partial^2 u}{\partial t^2}(t) + \mathfrak{K}u(t) = \mathfrak{B}b \cdot V(t), \tag{8.2}$$

at time $t \in \mathbb{R}^+$ over the domain $\Omega \subset \mathbb{R}^d$ and also satisfies the initial conditions

$$u(0) = u_0, \quad \text{and} \quad \frac{\partial u}{\partial t}(0) = v_0,$$

and problem-specific boundary conditions. In these initial examples the mass distribution operator $\mathfrak{M}$ is a multiplication operator $\mathfrak{M}(\cdot) = \rho_m A(\cdot)$, the external load function is $f(t,\cdot) = \mathfrak{B}bV(t)$ at time $t \in \mathbb{R}^+$, $\mathfrak{B}b$ is the control influence operator that is defined in terms of a spatial differential operator $\mathfrak{B}$ and a given spatial function b, and $V(t)$ is the input voltage at time $t \in \mathbb{R}^+$. We will see that the spatial differential operator $\mathfrak{K}$ is of order $2r$ and its action on a function u can be written using the summation convention over $0 \le |\alpha|, |\beta| \le r$ as

$$(\mathfrak{K}u)(x) := (-1)^{|\alpha|}\frac{\partial^{|\alpha|}}{\partial x^\alpha}\left(a_{\alpha\beta}(x)\frac{\partial^{|\beta|}u}{\partial x^\beta}(x)\right)$$

for some given spatial functions $a_{\alpha\beta}$, where $\alpha := (\alpha_1, \dots, \alpha_d)$ and $\beta := (\beta_1, \dots, \beta_d)$ are multi-indices of positive integers. The magnitude of a multi-index is defined as $|\alpha| := \sum_{i=1}^d \alpha_i$, and the partial differential operator $\frac{\partial^{|\alpha|}}{\partial x^\alpha}(\cdot)$ is defined by the expression

$$\frac{\partial^{|\alpha|}}{\partial x^\alpha}(\cdot) := \frac{\partial^{|\alpha|}}{\partial x_1^{\alpha_1}\cdots\partial x_d^{\alpha_d}}(\cdot).$$

In the examples considered in this chapter, the action of the spatial differential operator $\mathfrak{B}$ on a function w can be written using the summation convention over $0 \leq |\gamma| \leq r$ as

$$(\mathfrak{B}w)(x) := b_\gamma(x)\frac{\partial^{|\gamma|}}{\partial x^\gamma}w(x)$$

for some given functions b_γ and multi-index γ. We review the two simple examples for the composite piezoelectric rod and beam in Example 8.1.1.

Example 8.1.1 Consider the axial piezoelectric composite in Equation 7.21, but modified here so that the electrodes are located at $x_3 = a$ and $x_3 = b$. It is evident from Equation 6.11 that the mapping $\mathfrak{K}$ is the second order, spatial differential operator that acts on $u_3 := u_3(t, x_3)$ via the equation

$$\mathfrak{K}u_3 = -\frac{\partial}{\partial x_3}\left(AC_{33}^E\frac{\partial u_3}{\partial x_3}\right),$$

the spatial differential operator $\mathfrak{B}(\cdot) := \frac{\partial}{\partial x_3}(\cdot)$, $b := \frac{Ae_{33}}{L}\chi_{[a,b]}$, and the control influence operator is defined so that

$$\mathfrak{B}b \cdot V := \frac{\partial}{\partial x_3}\left(\frac{Ae_{33}}{L}\chi_{[a,b]}\right)V.$$

In the case of the piezoelectrically actuated composite beam, we see from Equation 6.20 that the mapping $\mathfrak{K}$ that acts on $u_3 := u_3(t, x_1)$ is given by the fourth order, spatial differential operator

$$\mathfrak{K}u_3 = \frac{\partial^2}{\partial x_1^2}\left(C_{11}^E I\frac{\partial^2 u_3}{\partial x_1^2}\right),$$

the differential operator $\mathfrak{B}(\cdot) := \frac{\partial^2}{\partial x_1^2}(\cdot)$, the function $b := \frac{Ke_{31}}{t_p}\chi_{[a,b]}$, and the control influence operator takes the form

$$\mathfrak{B}b \cdot V := \frac{\partial^2}{\partial x_1^2}\left(\frac{Ke_{31}}{t_p}\chi_{[a,b]}\right)V.$$

In both cases the zero order differential operator $\mathfrak{M}$, or multiplication operator, is given as $\mathfrak{M}(\cdot) = \rho_m A(\cdot)$.

These examples have identified the mass distribution multiplication operators and control influence operators for the axial and bending piezoelectric composites. They illustrate that the general Equation 8.2 is applicable to these cases. It is also important to observe that the approach developed here can be generalized with some minor modifications to include unknown displacement vectors and vectors of inputs. Reference [4] generalizes the class of systems beyond the simple models discussed in this book: it defines and analyzes the differential operators and control influence operators for linear and nonlinear plates, curvilinear plates and shells, coupled structural-acoustic problems, and coupled fluid-structural problems. Equation 8.2, however, can be viewed as a starting point in developing these more general models for such complex systems.

8.1.1 Classical Solutions

Different types of solutions of the governing equations can be defined by interpreting the derivatives that appear in Equation 8.2 as either classical or weak derivatives. See Appendix S.3 for a brief discussion of these different types of derivatives or consult a standard textbook on partial differential equations such as [40] or [46] for a rigorous and general presentation. We say that a function $u = u(t, x)$ is a *classical solution* of the governing equations if for each time $t \in \mathbb{R}^+$ and $x \in \Omega$ the function u satisfies Equations 8.2 where each spatial derivative that appears in the differential operators $\mathfrak{K}$ and $\mathfrak{B}$ is interpreted in the classical sense. The requirement that all the spatial derivatives that appear in Equation 8.2 are interpreted as classical derivatives is very stringent. It is often difficult or even impossible to justify. For example, in composite structures it is often the case that material properties are discontinuous. The axial stiffness AC_{33}^E or the bending stiffness $C_{11}^E I$ in many important cases may be a piecewise constant function. When this is the case, since a discontinuous function is not classically differentiable at its points of discontinuity, it is not possible to interpret the spatial differential operator $\mathfrak{K}$ in a classical sense at all points of the domain $\Omega \subset \mathbb{R}^d$. Similarly, a term such as $\frac{\partial^2}{\partial x_1^2}(\chi_{[a,b]})$ that appears in the control influence operator likewise cannot be interpreted in a classical sense since the characteristic function $\chi_{[a,b]}$ is not classically differentiable at $x_1 = a$, nor at $x_1 = b$.

Another technical reason that mathematicians and analysts have introduced different notions of a solution, in contrast to the classical solution, is that the alternatives can be more amenable to powerful operator-theoretic results for establishing well-posedness of problems. In particular, Hilbert space methods, such as can be found in [40, 46], or [41] have emerged over the past few decades as one of the standard approaches for studying partial differential equations. The weak formulation of the governing equations, discussed in some detail below, has become popular precisely because of its rich theoretical foundation for operators acting on Hilbert or Banach spaces. A wide variety of rigorous theorems have been derived for weak solutions of partial differential equations, and these results form the theoretical backbone of much contemporary analysis.

In addition to these theoretical considerations, there are pragmatic reasons that motivate the construction of approximations using alternative, weak solutions.

Suppose we seek an approximation of the classical interpretation of the governing equations directly. It is then necessary to construct the approximate solution so that it has all the requisite classical spatial derivatives and such that it satisfies the appropriate boundary conditions. Such an approximate solution must have two spatial derivatives for an axial rod, or four spatial derivatives for a beam or plate model. As the number of required spatial derivatives increases, so does the complexity and difficulty of generating a suitable approximation that satisfies the required boundary conditions. This task becomes even more difficult, and often intractable, when constructing classical solutions for linearly piezoelectric continua on irregular domains.

8.1.2 Strong and Weak Solutions

For all of these reasons, the alternative definitions of a solution have been defined that are not as restrictive as the classical solution. These definitions are based on interpreting the spatial derivatives that appear in the governing equations as *distributional derivatives* or *weak derivatives*. A function $u : \Omega \rightarrow \mathbb{R}$ is said to have a weak derivative v provided that

$$\int_{\Omega} v(x)\phi(x)dx = -\int_{\Omega} u(x)\frac{\partial \phi}{\partial x}(x)dx$$

for all smooth functions ϕ that have compact support in the domain Ω. In other words, a function u has a weak derivative if it satisfies an integration by parts formula when multiplied by any smooth function that vanishes in the neighborhood of the boundary of Ω. It is easy to see that any classically differentiable function is weakly differentiable, and in this case the weak derivative is equal to the classical derivative. Also, any function that is piecewise classically differentiable except at a finite number of points is weakly differentiable. Thus, while a piecewise constant function that represents the material properties is not classically differentiable, it is weakly differentiable.

A function u is a *strong solution* of the governing Equation 8.2 if all of the $2r$ derivatives that appear in the differential operator $\mathfrak{K}$ are interpreted as weak derivatives and each of these weak derivatives is square integrable. We denote by $L^2(\Omega)$ the collection of square integrable, real-valued functions over a domain $\Omega \subset \mathbb{R}^d$. In other words, we set

$$L^2(\Omega) := \left\{ f : \Omega \rightarrow \mathbb{R} | \; \| f \|^2_{L^2(\Omega)} := (f, f)_{L^2(\Omega)} := \int_D f^2 dv < \infty \right\}.$$

In this definition $\| f \|_{L^2(\Omega)}$ and $(f, g)_{L^2(\Omega)}$ are the norm of the function f and the inner product of functions f, g, respectively, in the space $L^2(\Omega)$. As discussed more fully in Appendix S.3, a function u is the strong solution of the governing Equations 8.2 provided that all the spatial derivatives in the equations are interpreted as weak derivatives and $u \in H^{2r}(\Omega)$ where $H^{2r}(\Omega)$ is the Sobolev space of functions in $L^2(\Omega)$ that have square integrable weak derivatives of order less than or equal to $2r$. That is, we have $u \in H^{2r}(\Omega)$ with

$$H^{2r}(\Omega) := \left\{ f \in L^2(\Omega) \mid \frac{\partial^{|\alpha|} f}{\partial x^\alpha} \in L^2(\Omega) \quad \forall |\alpha| \leq 2r \right\}.$$

Equations 8.1 and 8.2 are consequently sometimes known as the *strong form* of governing equations. See [40] for a rigorous statement of this definition.

Even though the strong solution is more general than the classical solution, it can be difficult to construct approximate solutions that have the requisite $2r$ weak derivatives and that also satisfy the boundary conditions. For purposes of building approximate solutions, it can be most convenient to consider a notion of a solution that is even more general, the *weak solution*. A weak solution u is a solution of the weak form of the governing equations, and it need only have r weak derivatives that are square integrable, that is $u \in H^r(\Omega)$. The first step in defining the *weak form* of the governing equations takes the inner product $(\cdot, \cdot)_{L^2(\Omega)}$ in the set of square integrable functions $L^2(\Omega)$ of Equation 8.2 with an arbitrary, sufficiently smooth function v. For any choice of such a function v, the displacement $u(t) := u(t, \cdot)$ obeys the relation

$$\left(\rho_m A \frac{\partial^2 u}{\partial t^2}(t), v \right)_{L^2(\Omega)} + (\mathfrak{K} u(t), v)_{L^2(\Omega)} = (\mathfrak{B} b \cdot V(t), v)_{L^2(\Omega)},$$

at time $t \in \mathbb{R}^+$. This equation can also be written as the integral equation

$$\int_\Omega \rho_m A \frac{\partial^2 u}{\partial t^2}(t) v dx + \int_\Omega \mathfrak{K} u(t) v dx = \int_\Omega \mathfrak{B} b \cdot V(t) v dx.$$

The function v in these equations is often referred to as a *test function* or *trial function*. The spatial derivatives that appear in the inner product $(\mathfrak{K} u(t), v)_{L^2(D)}$ are then integrated by parts to move half of the derivatives that define the even order operator $\mathfrak{K}$ from u onto v. Similarly, the control influence terms are integrated by parts to move all of the derivatives in $\mathfrak{B}$ that act on b onto v. Upon integrating by parts, we obtain the *weak form* of the governing equations which seeks a solution $u(t) := u(t, \cdot)$ that satisfies

$$\left(\rho_m A \frac{\partial^2 u}{\partial t^2}(t), v \right)_{L^2(\Omega)} + a(u(t), v) + BT = (\mathfrak{B}^* v, bV(t))_{L^2(\Omega)}. \tag{8.3}$$

for all admissible trial functions v, where $a(\cdot, \cdot)$ is a symmetric non-negative bilinear form, $\mathfrak{B}^*$ is the formal adjoint of $\mathfrak{B}$ defined as

$$(\mathfrak{B}^* w)(x) := (-1)^{|\gamma|} \frac{\partial^{|\gamma|}}{\partial x^\gamma} (b_\gamma(x) w(x)),$$

and all the boundary terms generated during the integration by parts are collected in BT. Note that the formal adjoint $\mathfrak{B}^*$ is defined so that $(\mathfrak{B} w, v)_{L^2(\Omega)} = (w, \mathfrak{B}^* v)_{L^2(\Omega)}$ for all smooth functions w, v that have compact support in Ω.

The next few examples show how this process of constructing the weak form of the governing equations proceeds in typical cases.

Example 8.1.2 (Uniaxial Rod). In this example, we modify the axial composite so that the piezoelectric material is located in the subdomain $[a,b] \subset [0,L] = \Omega \subset \mathbb{R}$. The material in the complementary region $[0,a) \cup (b,L]$ is assumed to be linearly elastic so that $e_{33} = 0$ for $x_3 \notin [a,b]$. The weak form of the governing equations for the linearly piezoelectric rod is obtained by multiplying the strong form of the governing Equation 6.11 by an arbitrary function $v = v(x_3)$ and integrating by parts in the spatial variable x_3 to move one spatial derivative from u_3 onto v. We likewise integrate the control influence operator to move one spatial derivative onto v. In the weak formulation, then, we seek a function $u_3(t) := u_3(t,\cdot)$ that satisfies

$$\int_0^L \rho_m A \frac{\partial^2 u_3}{\partial t^2}(t) v dx_3 + \int_0^L A C_{33}^E \frac{\partial u_3}{\partial x^3}(t) \frac{\partial v}{\partial x_3} dx_3 \tag{8.4}$$

$$-A\left(C_{33}^E \frac{\partial u_3}{\partial x_3}(t) + \frac{e_{33}\chi_{[a,b]}}{b-a} V(t) \right) v \Big|_0^L$$

$$= -\int_0^L \frac{A e_{33}\chi_{[a,b]}}{b-a} \frac{\partial v}{\partial x_3} dx_3 V(t) \tag{8.5}$$

for all admissible choices of the test function $v = v(x_3)$. Carefully note that the integration by parts yields the boundary terms $-AT_{33}v|_0^L$ in the second line in Equation 8.5. One should compare this boundary term to the variational boundary condition for the axial rod derived in Section 7.3.1. The variational boundary condition in Section 7.3.1 has the same form as the boundary term in Equation 8.5 if we identify $v \equiv \delta u_3$. If the end of the rod at $x_3 = 0$ or $x_3 = L$ is subject to a homogeneous Dirichlet boundary condition, then we define the admissible functions v so that $v(0) = 0$ or $v(L) = 0$, respectively. These conditions define the set of admissible test functions. If the end of the rod at $x_3 = 0$ or $x_3 = L$ is free, then we set $AT_{33}(0) = 0$ or $AT_{33}(L) = 0$, respectively. In either case, when the boundary conditions on u_3 are enforced and v is an admissible test function, the second line in Equation 8.5 is equal to zero. The governing equations reduce to that in the first line. In Equation 8.3, the stiffness bilinear form $a(\cdot,\cdot)$ is consequently defined as

$$a(u_3(t), v) := \int_0^L A C_{33}^E \frac{\partial u_3}{\partial x_3}(t) \frac{\partial v}{\partial x_3} dx_3.$$

(Continued)

Example 8.1.2 (Continued)

The formal adjoint $\mathfrak{B}^*$ is given by

$$\mathfrak{B}^* v := -\frac{\partial v}{\partial x_3}.$$

We see that the weak form of the governing equations has the structure shown in Equation 8.3 with $BT = 0$.

Example 8.1.3 (Bernoulli–Euler Beam). The weak form of the equations governing the piezoelectrically actuated composite Bernoulli–Euler beam are obtained in a manner similar to that outlined in Example 8.1.2. We multiply the strong form of the governing equations by an arbitrary test function $v = v(x_1)$ and subsequently integrate by parts twice to shift two spatial derivatives from the unknown displacement u onto the test function v. We also integrate the control influence operator by parts twice to move two spatial derivatives onto v. The weak solution $u_3(t) = u_3(t, \cdot)$ must therefore satisfy the equation

$$\int_0^L \rho_m A \frac{\partial^2 u_3}{\partial t^2}(t) v dx_1 + \int_0^L C_{11}^E I \frac{\partial^2 u_3}{\partial x_1^2}(t) \frac{\partial^2 v}{\partial x_1^2} dx$$

$$= \int_0^L \frac{K e_{31} \chi_{[a,b]}}{t_p} \frac{\partial^2 v}{\partial x_1^2} dx_1 V(t) \tag{8.6}$$

$$- \frac{\partial}{\partial x_1} \left(-C_{11}^E I \frac{\partial^2 u_3}{\partial x_1^2}(t) + \frac{K e_{31} \chi_{[a,b]}}{t_p} V(t) \right) v \Bigg|_0^L \tag{8.7}$$

$$+ \left(-C_{11}^E I \frac{\partial^2 u_3}{\partial x_1^2}(t) + \frac{K e_{31} \chi_{[a,b]}}{t_p} V(t) \right) \frac{\partial v}{\partial x_1} \Bigg|_0^L \tag{8.8}$$

for all admissible test functions v. Again, one should compare the boundary terms in the weak form of the governing equations in lines 8.7 and 8.8 to the variational boundary conditions for the Bernoulli–Euler beam in Section 7.3.2. The two expressions are identical when we interpret $v \equiv \delta u_3$ and $\frac{\partial v}{\partial x_1} \equiv \delta \left(\frac{\partial u_3}{\partial x_1} \right) \equiv \frac{\partial(\delta u_3)}{\partial x_1}$. If the displacement of the ends of the beam are constrained to be zero at $x_1 = 0$ or $x_1 = L$, then the test function v is selected so that $v(0) = 0$ or $v(L) = 0$, respectively. If the slopes of the ends of the beam are constrained to be zero at $x_1 = 0$ or $x_1 = L$, then the slope of the test functions are chosen to satisfy $\frac{\partial v}{\partial x_1}(0) = 0$ or $\frac{\partial v}{\partial x_1}(L) = 0$, respectively. It should be recognized that the boundary terms BT in Equations 8.7 and 8.8

are just

$$BT = -\mathcal{S}v\Big|_0^L + \mathcal{M}\frac{\partial v}{\partial x_1}\Big|_0^L,$$

with $\mathcal{S}$ and $\mathcal{M}$ the transverse shear force and bending moment, respectively, in the beam. If the ends of the beam are free, then the shear force and bending moment are set equal to zero. In all cases, the boundary terms that appear in the second and third lines in Equations 8.7 and 8.8 drop out of the weak form of the equations for all admissible choices of the trial functions v. In this case the bilinear form $a(\cdot,\cdot)$ is defined as

$$a(u_3(t), v) := \int_0^L C_{11}^E I \frac{\partial^2 u_3}{\partial x_1^2}(t)\frac{\partial^2 v}{\partial x_1^2}dx,$$

and the formal adjoint $\mathfrak{B}^*$ is given by

$$\mathfrak{B}^* v := \frac{\partial^2 v}{\partial x_1^2}.$$

It is evident that the weak form of the equations again have the structure shown in Equation 8.3 with the contribution of boundary terms $BT = 0$.

8.2 Modeling Damping and Dissipation

The strong form of the governing Equation 8.2 and the weak form in Equation 8.3 make no explicit provision for structural energy dissipation and damping. The bilinear form $a(u(t), v)$ ordinarily represents stored structural or strain energy, inertial terms are contained in the term $(\rho_m A\frac{\partial^2 u}{\partial t^2}(t), v)_{L^2(\Omega)}$, and the work done by the control input is implicit in the term $(\mathfrak{B}b \cdot V(t), v)_{L^2(\Omega)}$ in Equation 8.3. Modeling the energy dissipation mechanisms in structural problems can be notoriously difficult [11, 21]. Likewise, it can be challenging to make experimental measurements that characterize these contributions to the equations of motion. Generally speaking, damping models can be introduced in the MDOF systems that result from approximating a DPS, or they can be introduced directly into the DPS. One can find examples of damping models for DPS or evolutionary partial differential equations in [4] or [7]. Descriptions of damping models that can be incorporated directly in SDOF or MDOF systems can be found in [21] or [11].

One particular difficulty that may be encountered in constructing such damping models is that it is often not simple to "upscale" a local energy dissipation property at a fine scale to the macroscale to obtain the structural response. There are a variety of methods that can be used in principle to measure the local energy dissipation, for example, by characterizing vibration decay rates of local test specimens. Unfortunately, micromechanical measurements of energy dissipation in materials can be ill-suited for extension to the entire structure. In contrast, it is well documented that local mass density properties of a material can be used to determine the mass matrix, which provides an important tool for determining the system response at the structural scale. Similarly, it is standard to measure local stiffness properties and use them to create a stiffness matrix, which is again essential for modeling the response at the structural level. However, for many composite structures, and certainly for many of the models in this text, such is not the case for energy dissipation. If a structure is built from many constituent substructures that are bonded together or assembled using common joint architectures, the energy dissipation on the overall structural scale of a composite can be a delicate and complex function of the how the joints are fabricated. This tends to be true for active material composites that are constructed on a benchtop in a laboratory, for example, where bonding layers may be poorly understood or characterized. They may vary from one specimen to the next during fabrication. Small variations in joint properties can manifest as large differences in structural energy dissipation even though measurements of the structural constants of the specimens are close to the same. On the other hand, a single crystal piezoelectric specimen is one example of a system in which local damping measurements may be amenable to upscale to determine global structural energy dissipation. This special case, however, is not representative of many of the composite structures designed to serve as actuators in this text.

For these reasons, it has become a standard practice to include a structural energy dissipation model in the governing equations that is based on structural scale phenomenology, rather than one based on some micromechanical model or measurement. Many such models have been documented in the vibrations literature. See [21] or [11], for example, for discussions of the viscous damping and Coulomb damping models in engineering texts. These models are used primarily for SDOF or MDOF models. Perhaps the most common, and certainly the most common for finite dimensional systems discussed in [21] or [11], is the viscous damping model.

When we seek to enhance a classical or strong formulation of the governing equations to account for macroscopic energy loss, an additional term is introduced into the governing equations, Equation 8.1 depicts a modified form of the general equations that includes the damping operator $\mathfrak{C}$. Alternatively, a solution $u(t) := u(t,\cdot)$ of the weak form of the governing equations with viscous damping satisfies the equation

$$\left(\rho_m A\frac{\partial^2 u}{\partial t^2}, v\right)_{L^2(\Omega)} + a_1\left(\frac{\partial u}{\partial t}, v\right) + a(u,v) + BT = (\mathsf{B}^* v, bV)_{L^2(\Omega)} \tag{8.9}$$

for any admissible test function v. All of the terms that appear in 8.9 have been defined in Equation 8.3, but in addition the energy dissipation is represented by the bounded, symmetric, bilinear form $a_1(\cdot,\cdot)$. It is sometimes possible to relate the damping operator $\mathfrak{C}$ in Equation 8.1 to the bilinear form $a_1(\cdot,\cdot)$ that represents damping in the weak Equation 8.9.

Example 8.2.1 In this example we construct a highly simplified case that illustrates how a bounded, linear damping operator $\mathfrak{C} : H \rightarrow H$ can be related to a bounded, damping bilinear form in the special case that $a_1(\cdot,\cdot) : H \rightarrow H$ where H is a Hilbert Space. It is then possible to define the bounded linear operator $\mathfrak{C}$ in terms of the damping bilinear form from the identity that $(\mathfrak{C}u, v)_H := a_1(u, v)$ for all $u, v \in H$. The converse is also true, and it is possible to define the damping bilinear form $a_1(\cdot,\cdot)$ in terms of the operator $\mathfrak{C}$ using the same formula. This is easily checked if it happens that the Hilbert space is actually finite dimensional, $H = \mathbb{R}^d$ for some finite $d > 0$. This rough strategy is also popular for defining damping in formulations for which a set of eigenmodes $\{\psi_j\}_{j=1}^{\infty}$ have been derived that are dense in H, as discussed in Appendix S.1.2. Then we can define $\mathfrak{C}$ by the condition that $(\mathfrak{C}u, v)_H := \sum_{i,j=1}^{\infty} a_1(\psi_i, \psi_j)u_i v_j$ for all $u, v = H$ that have the representations $u = \sum_{j=1}^{\infty} u_j\psi_j$ and $v = \sum_{j=1}^{\infty} v_j\psi_j$, respectively.

It should be noted that this example, while it does provide motivation and intuition, is not at all the most general situation. The full generality of the weak solution often chooses to define the damping bilinear form $a_1 : V \times V \rightarrow \mathbb{R}$ for a Banach space $V \subseteq H$. The proper interpretation of the weak solution then is made precise in terms of a Gelfand triple $V \subseteq H \equiv H^* \subseteq V^*$ where V^* and H^* are the topological duals of V and H, respectively. The technical details attendant to this more general analysis far exceeds the scope of this text. See [4, 7] for a detailed discussion.

8.3 Galerkin Approximations

Galerkin methods build approximations of the unknown solution u of the weak Equations 8.3 as a linear combination of a finite number n_d of basis functions $\{N_j\}_{j=1}^{n_d}$. The integer $n_d > 0$ is the number of degrees of freedom (DOF) of the approximate system. Each of the basis functions N_j for $j = 1, \ldots, n_d$ is defined over the spatial domain Ω. Again using the summation convention we write

$$u(t,x) := N_j(x)u_j(t) \tag{8.10}$$

where each $u_j(t)$ is an unknown function of time $t \in \mathbb{R}^+$. The $u_j(t)$ are the DOF variables for the multi-input, multi-output (MIMO) system that results from Galerkin approximation of the DPS. The MIMO system of n_d ordinary differential equations results when we substitute the Galerkin approximation in Equation 8.10 into the weak form of the Equations 8.3 and then choose the trial function to be $v = N_i$. We thereby obtain one equation

$$\int_{\Omega} \rho_m A N_i N_j dx \ddot{u}_j(t) + a(N_i, N_j) u_j(t) + BT = \int_{\Omega} \mathfrak{B}^* N_i b V(t) dx, \tag{8.11}$$

with the summation convention over $j = 1, \ldots, n_d$ for each choice of the trial function $v = N_i$, $i = 1, \ldots, n_d$. The resulting equations can be collected in a matrix form as follows:

$$\begin{bmatrix} \int_{\Omega}\rho_m A N_1 N_1 dx & \cdots & \int_{\Omega}\rho_m A N_1 N_{n_d} dx \\ \vdots & \ddots & \vdots \\ \int_{\Omega}\rho_m A N_{n_d} N_1 dx & \cdots & \int_{\Omega}\rho_m A N_{n_d} N_{n_d} dx \end{bmatrix} \begin{Bmatrix} \ddot{u}_1 \\ \vdots \\ \ddot{u}_{n_d} \end{Bmatrix}$$

$$+ \begin{bmatrix} a(N_1, N_1) & \cdots & a(N_1, N_{n_d}) \\ \vdots & \ddots & \vdots \\ a(N_{n_d}, N_1) & \cdots & a(N_{n_d}, N_{n_d}) \end{bmatrix} \begin{Bmatrix} u_1 \\ \vdots \\ u_{n_d} \end{Bmatrix} + BT$$

$$= \begin{Bmatrix} \int_{\Omega} b \mathfrak{B}^* N_1 dx \\ \vdots \\ \int_{\Omega} b \mathfrak{B}^* N_{n_d} dx \end{Bmatrix} V(t).$$

These equations can be written using the summation convention over $j = 1, \ldots, n_d$ as

$$m_{ij} \ddot{u}_j(t) + k_{ij} u_j(t) + BT = b_i V(t),$$

for each $i = 1, \ldots, n_d$, or they can be written succinctly in matrix form as

$$\boldsymbol{m}\boldsymbol{u}(t) + \boldsymbol{k}\boldsymbol{u}(t) + BT = \boldsymbol{b}V(t)$$

with the mass matrix $\boldsymbol{m} \in \mathbb{R}^{n_d \times n_d}$, the stiffness matrix $\boldsymbol{k} \in \mathbb{R}^{n_d \times n_d}$, the control influence matrix $\boldsymbol{b} \in \mathbb{R}^{n_d \times 1}$, and the n_d vector of DOFs $\boldsymbol{u} \in \mathbb{R}^{n_d \times 1}$. As noted above, when all of the boundary conditions are enforced and v is restricted to be a admissible test function, it normally follows that $BT = 0$.

Galerkin approximation methods for undamped systems culminate in the degree of freedom Equations 8.11. For those structures that dissipate energy via viscous damping that is modeled in terms of a bilinear form $a_1(\cdot, \cdot)$ as in

Equation 8.9, the procedure is nearly identical. With the introduction of the Galerkin approximation $u(t,x) := \sum N_j(x)u_j(t)$ for $j = 1, \ldots, n_d$, the DOFs satisfy the finite set of equations

$$m_{ij}\ddot{u}_j(t) + c_{ij}\dot{u}_j(t) + k_{ij}u_j(t) + BT = b_iV(t)$$

for $i = 1, \ldots, n_d$ when the system is viscously damped. The mass matrix m_{ij}, stiffness matrix k_{ij}, control influence matrix b_i, and boundary terms BT are defined as in Equation 8.11. The damping matrix c_{ij} is defined in terms of the bilinear form $a_1(\cdot,\cdot)$,

$$c_{ij} = a_1(N_i, N_j).$$

Specific forms of the equations generated for the Galerkin approximation of the linearly piezoelectric axial rod and Bernoulli–Euler beam are presented in Examples 8.3.1 and 8.3.2.

Example 8.3.1 (Uniaxial Rod). We consider again the linearly piezoelectric composite axial rod from Equations 8.4 and 8.5 and build a Galerkin approximation of its solution. We modify the problem here slightly and assume that the piezoelectric material is confined to the region $[a,b] \subset [0,L] = \Omega$. For each $x \in [x,a) \cup (b,L]$, the material is assumed to be linearly elastic. Let us assume that the linearly piezoelectric rod is constrained so that it is fixed at $x_3 = 0$ and free at $x_3 = L$. We approximate the weak solution as the sum

$$u_3(t,x_3) \approx N_j(x_3)u_{3,j}(t)$$

in Equations 8.4 and 8.5 where each N_j is a shape function of the spatial variable $x_3 \in [0,L] = \Omega \subset \mathbb{R}^1$. When we introduce the test function v in Equations 8.4 and 8.5 and integrate by parts, we obtain the Galerkin approximation Equations 8.11. The shape functions are assumed to satisfy two critical properties: (1) each function N_j must be sufficiently smooth so that the all the integrals in the weak form of the equations exist and are finite, and (2) the shape functions N_j must be selected so that u_3 satisfies all the geometric boundary conditions. It is evident from inspection of the weak form of the equations that if N_j and its derivative $\frac{dN_j}{dx_3}$ are square integrable for $j = 1, \ldots, n_d$, then each coefficient in the weak solution is computable. The table below summarizes the two cases of boundary conditions on u considered in this example and the corresponding boundary conditions on v that restrict and define the set of admissible test functions.

(Continued)

Example 8.3.1 (Continued)

Case	BC on $u_3(t, x_3)$	BC on N_i	BC on v
1	$u_3(t,0) = u_3(t,L) = 0$	$N_j(0) = N_j(L) = 0$	$v(0) = 0, v(L) = 0$
2	$u_3(t,0) = T_{33}(t,L) = 0$	$N_i(0) = \frac{dN_i}{dx_3}(L) = 0$	$v(0) = 0, \frac{dv}{dx_3}(L)$

We first discuss Case 1 in the table. If we choose each N_j so that it satisfies the homogeneous boundary conditions $N_j(0) = 0$ and $N_j(0) = 0$, the approximate solution $u_3(t, x_3)$ also satisfies the kinematic boundary conditions $u_3(t, 0) = 0$ and $u_3(t, L) = 0$. We also require in Case 1 that all the admissible trial functions satisfy $v(0) = 0$ and $v(L) = 0$. It is evident that we can then choose $v = N_i$ for $i = 1, \ldots, n_d$ in Equations 8.4 and 8.5. Both boundary terms in the second line in Equation 8.5 vanish and therefore $BT = 0$. Now we turn to the Case 2 in the table. When we choose $N_j(0) = 0$, we again guarantee that $u_3(t, 0) = 0$ as required. If we additionally require that $\frac{dN_j}{dx_3}(L) = 0$, we ensure that $T_{33}(t, L) = 0$. If we choose the test function $v = N_i$ for $i = 1, \ldots, n_d$ in Equations 8.4 and 8.5, we thereby guarantee that $v(0) = 0$ and $\frac{dv}{dx_3}(L) = 0$. On inspecting the expression in line 8.5, we see that the boundary terms vanish and $BT = 0$ in Equation 8.11. We write Equation 8.11 choosing $v = N_i$ for $i = 1, \ldots, n_d$ and obtain the set of equations

$$\int_0^L \rho_m A N_i N_j dx_3 \frac{d^2 u_{3,j}}{dt^2}(t) + \int_0^L A C_{33}^E \frac{dN_i}{dx_3}\frac{dN_j}{dx_3} dx_3 u_{3,j}(t)$$
$$= -\int_0^L \frac{A e_{33} \chi_{[a,b]}}{b-a} \frac{dN_i}{dx_3} dx_3 V(t).$$

It is conventional to introduce the mass matrix m_{ij}, the stiffness matrix k_{ij}, and the control influence matrix b_i as

$$m_{ij} = \int_0^L \rho_m A N_i N_j dx_3,$$
$$k_{ij} = \int_0^L A C_{33}^E \frac{dN_i}{dx_3}\frac{dN_j}{dx_3} dx_3,$$
$$b_i = -\int_0^L \frac{A e_{33} \chi_{[a,b]}}{b-a} \frac{dN_i}{dx_3} dx_3,$$

and obtain the final form of the equations that are satisfied by the Galerkin approximation of the weak equations,

$$m_{ij} \frac{d^2 u_{3,j}}{dt^2}(t) + k_{ij} u_{3,j}(t) = b_i V(t),$$

with the summation convention over $j = 1, \dots, n_d$. There are many possible choices for the shape functions $\{N_i\}_{i\in\mathbb{N}}$ in this construction. The most common choices are eigenfunctions or finite element shape functions. We briefly discuss each choice in Sections 8.3.1 and 8.3.2.

Example 8.3.2 (Galerkin Approximation, Beam). The Galerkin approximation of the the piezoelectric composite Bernoulli–Euler beam represents the displacement $u_3 = u_3(x_1)$ as the sum

$$u_3(t, x_1) \approx N_j(x_1) u_{3,j}(t),$$

with the summation convention over $j = 1, \dots, n_d$. When Equations 8.6, 8.7, and 8.8 are written for the choice of trial function $v := N_i$ for $i = 1, \dots, n_d$, we obtain the set of ordinary differential equations

$$\int_0^L \rho_m A N_i N_j dx_1 \frac{\partial^2 u_{3,j}}{\partial t^2}(t) + \int_0^L C_{11}^E I \frac{d^2 N_i}{dx_1^2} \frac{d^2 N_j}{dx_1^2} dx_1 u_{3,j}(t)$$

$$= \int_0^L \frac{K e_{31} \chi_{[a,b]}}{t_p} \frac{d^2 N_i}{dx_1^2} dx_1 V(t), \tag{8.12}$$

$$-\frac{\partial}{\partial x_1}\left(-C_{11}^E I \frac{\partial^2 u_3}{\partial x_1^2}(t) + \frac{K e_{31} \chi_{[a,b]}}{t_p} V(t) \right) N_i |_0^L, \tag{8.13}$$

$$+\left(-C_{11}^E I \frac{\partial^2 u_3}{\partial x_1^2}(t) + \frac{K e_{31} \chi_{[a,b]}}{t_p} V(t) \right) \frac{dN_i}{dx_1} |_0^L. \tag{8.14}$$

Carefully note that the boundary terms BT in Equations 8.13 and 8.14 can be summarized as

$$BT = -S N_i \Big|_0^L + \mathcal{M} \frac{\partial N_i}{\partial x_1}\Big|_0^L,$$

with S the transverse shear force and $\mathcal{M}$ the bending moment at a location $x_1 \in [0, L]$. Again, the shape functions $N_i = N_i(x_1)$ must be selected so that (1) all the integrals in the weak form of the equations make sense, and (2) the approximate solution satisfies the geometric boundary conditions. Inspection of the terms that make up the weak form of the equations shows that if the shape functions N_i and their second derivatives $\frac{\partial^2 N_i}{\partial x_1^2}$ are square integrable, then all the integral terms can be computed. The appropriate conditions on the admissible trial functions N_i can again be determined by comparing the boundary terms in the second and third lines in Equations 8.13 and

(Continued)

Example 8.3.2 (Continued)

8.14 to the variational boundary conditions in Section 7.3.2. These boundary expressions are the same if we identify $N_i := \delta u_3$ and $\frac{\partial N_i}{\partial x_1} := \delta \frac{\partial u_3}{\partial x_1} \equiv \frac{\partial(\delta u_3)}{\partial x_1}$. We summarize the boundary conditions for a clamped-free beam in the table below.

BC on $u_3(t, x_1)$	BC on N_i	BC on v
$u_3(t,0) = \frac{\partial u_3}{\partial x_1}(0) = 0$	$N_i(0) = \frac{dN_i}{dx_1}(0) = 0$	$v(0) = \frac{dv}{dx_3}(0) = 0$
$\mathcal{M}(L) = \mathcal{S}(L) = 0$	$\frac{d^2N_i}{dx_1^2}(L) = \frac{d^3N_i}{dx_1^3}(L) = 0$	$\frac{d^2v}{dx_3^2}(L) = \frac{d^3v}{dx_1^3}(L) = 0$

When we choose $N_i(0) = \frac{dN_i}{dx_1}(0) = 0$, it is immediate that the displacement u satisfies the geometric boundary conditions $u_3(t,0) = \frac{du_3}{x_1}(t,L) = 0$. When we enforce the condition that $\frac{d^2N_i}{dx_1^2}(L) = \frac{d^3N_i}{dx_1^3}(L) = 0$, the bending moment and shear force satisfy the boundary conditions $\mathcal{M}(L) = \mathcal{S}(L) = 0$. Because the boundary conditions on v are the same as those on N_i, we can choose $v = N_i$ as an admissible test function. With these choices, $BT = 0$ in the approximation of the weak equations 8.6 through 8.8, and the final set of governing ordinary differential equations reduces to the collection

$$\int_0^L \rho_m A N_i N_j dx_1 \frac{\partial^2 u_{3,j}}{\partial t^2}(t) + \int_0^L C_{11}^E I \frac{d^2N_i}{dx_1^2}\frac{d^2N_j}{dx_1^2} dx_1 u_{3,j}(t) \tag{8.15}$$

$$= \int_0^L \frac{K e_{31} \chi_{[a,b]}}{t_p} \frac{d^2N_i}{dx_1^2} dx_1 V(t), \tag{8.16}$$

for $i = 1, \ldots, n_d$. We introduce the mass matrix m_{ij}, stiffness matrix k_{ij}, and control influence matrix b_i as follows:

$$m_{ij} := \int_0^L \rho_m A N_i N_j dx_1, \tag{8.17}$$

$$k_{ij} := \int_0^L C_{11}^E I \frac{d^2N_i}{dx_1^2}\frac{d^2N_j}{dx_1^2} dx_1, \tag{8.18}$$

$$b_i := \int_0^L \frac{K e_{31} \chi_{[a,b]}}{t_p} \frac{d^2N_i}{dx_1^2} dx_1. \tag{8.19}$$

The final form of the governing equations are then given in terms of the summation convention as

$$m_{ij} \frac{d^2 u_{3,j}}{dt^2}(t) + k_{ij} u_{3,j}(t) = b_i V(t).$$

Examples 8.3.1 and 8.3.2 have presented in general terms how to construct Galerkin approximations of the weak form of the governing Equations 8.3 for prototypical linearly piezoelectric structures. However, the choice of basis functions $\{N_i\}_{i=1}^{n_d}$ used in the Galerkin approximation in Equation 8.10 has not yet been discussed in detail. Sections 8.3.1 and 8.3.2 discuss the two most common methods for constructing the basis functions used in Galerkin approximations.

8.3.1 Modal or Eigenfunction Approximations

Linear systems whose strong or weak governing equations have the form shown in Equations 8.2 and 8.3, respectively, have been studied carefully over the years in fields such as vibrations, structural dynamics, or applied mathematics. The systems in Equations 8.2 or 8.3 are examples of distributed parameter systems. Of all the techniques that have been employed over the years to build approximations of distributed parameter systems, perhaps the first that is taught to new students uses eigenfunctions of the spatial operator as the approximating basis $\{N_j\}_{j=1}^{n_d}$. Entire chapters of standard textbooks on vibrations such as [29] or on structural dynamics [11] have been devoted to this approach. The overwhelming advantage of this strategy is that, when it is applicable, the finite set of resulting ordinary differential Equations 8.11 are uncoupled. In other words, the choice of eigenfunctions renders the mass and stiffness matrices diagonal. Because the system of equations are uncoupled, it is possible to derive a closed form, analytic expression for the approximate solution.

Unfortunately, there is a significant drawback to employing Galerkin approximations based on the construction eigenfunctions. In general, it is possible to solve for the eigenfunction basis in only the most simple problems: the material properties of the piezoelectric composite must be uniform and the domain must be regular or highly structured.

Approximations that are based on eigenfunctions can be developed by starting with the strong form of the governing Equations 8.2, written here again in the form

$$\rho_m A \frac{\partial^2 u}{\partial t^2}(t) + \mathfrak{K} u(t) = \mathfrak{B} b V(t).$$

It has been noted in Section 8.1 that the operator $\mathfrak{K}$ is a spatial operator of order $2r$ and the mass distribution multiplication operator is $\mathfrak{M}(\cdot) := \rho_m A(\cdot)$ in our example problems. Eigenfunction methods begin by solving the abstract eigenvalue problem associated with the operators $\mathfrak{K}$ and $\mathfrak{M}$. A *nontrivial* functions ψ defined on the spatial domain Ω is a generalized eigenfunction corresponding to the eigenvalue λ of the operators $\mathfrak{K}$ and $\mathfrak{M}$ if it satisfies the equation

$$(\mathfrak{K} + \lambda \mathfrak{M})\psi = 0, \tag{8.20}$$

subject to a specified set of boundary conditions that arise from the geometry of a particular problem. Appendix S.1.2 gives a brief overview of eigenvalue and eigenfunction problems and summarizes their common properties. For our class of problems, there will at most be a countably infinite number of eigenvalues. In addition, if the eigenvalues of two eigenfunctions are distinct, then the associated eigenfunctions satisfy orthogonality conditions discussed below. We first illustrate this property in the following example that uses the eigenfunctions associated with the linearly elastic axial rod as the basis for an approximation of the homogeneous solution of the weak Equation 8.16.

Example 8.3.3 When we assume that the material and structural parameters are spatially uniform, Equation 8.20 takes the form

$$-AC_{33}^{E}\frac{d^2\psi}{dx^2} + \lambda\rho_m A\psi = 0$$

for the axial rod, subject to the boundary conditions $\psi(0) = \psi(L) = 0$. The equation is often rewritten as

$$\frac{d^2\psi}{dx^2} = \hat{\lambda}\psi$$

with $\hat{\lambda} := \rho_m\lambda/C_{33}^{E}$. From Example S.1.3 in the Appendix, we find that their are a countably infinite number of eigenpairs $(\hat{\lambda}_j, \psi_j) := (-\omega_j^2, \psi_j) = (-(j\pi/L)^2, \sin(j\pi x/L))$ for $j = 1, \dots, \infty$ for this equation. It is demonstrated in Example S.1.3 that eigenfunctions ψ_j corresponding to distinct eigenvalues are orthogonal in $L^2(\Omega)$ with $\Omega := [0, L]$. In particular, we have

$$(\psi_i, \psi_j)_{L^2(\Omega)} := \int_0^L \sin\frac{i\pi x}{L}\sin\frac{j\pi x}{L}dx = \begin{cases} \frac{L}{2} & \text{if } i = j \\ 0 & \text{otherwise,} \end{cases}$$

when the boundary conditions are $u_3(t, 0) = 0$ and $u_3(t, L) = 0$. Since $\alpha\psi_j$ is an eigenfunction if and only if ψ_j is an eigenfunction, there are an infinite number of ways to normalize the eigenfunctions. In this example we simply choose $\psi_j := \sin(j\pi x/L)$, which results in ψ_j that are orthogonal, but they do not have $(\psi_j, \psi_j)_{L^2(\Omega)} = \| \psi_j \|_{L^2(\Omega)}^2 = 1$. The Appendix discusses the normalization $\psi_j := \sin(j\pi x/L) \Big/ \sqrt{\int_0^L \sin^2((j\pi\eta)/L)d\eta}$ that yields the orthonormal eigenfunctions that satisfy $(\psi_i, \psi_j)_{L^2(D)} = \delta_{ij}$. The method of modal truncation then seeks a Galerkin approximation of the homogeneous solution of the weak equations in the form

$$u_3(t, x) = \psi_j(x)u_{3,j}(t) \tag{8.21}$$

where the summation convention holds for $j = 1, \ldots, n_d$. With this choice the mass matrix takes the diagonal form

$$m_{ij} = \int_0^L \rho_m A N_i(x_3) N_j(x_3) dx_3 = \rho_m A \int_0^L \sin\frac{i\pi x_3}{L} \sin\frac{j\pi x_3}{L} dx_3$$

$$= \rho_m A (\psi_i, \psi_j)_{L^2(D)} = \begin{cases} 0 & \text{if } i \neq j, \\ m_i := \rho_m A \int_0^L \psi_i^2(x_1) dx_1 & \text{if } i = j. \end{cases}$$

The constant m_i is knows as the modal mass of the ith degree of freedom. With a bit of computation, it can also be shown that the stiffness matrix is also diagonal. By definition, we have

$$k_{ij} = \int_0^L A C_{33}^E \frac{dN_i}{dx_3} \frac{dN_j}{dx_3} dx_3 = A C_{33}^E \int_0^L A C_{33}^E \frac{d\psi_i}{dx_3} \frac{d\psi_j}{dx_3} dx_3,$$

$$= A C_{33}^E \left\{ \frac{d\psi_i}{dx_3} \psi_j \Big|_0^L - \int_0^L \psi_j \frac{d^2\psi_i}{dx_3^2} dx_3 \right\} = -A C_{33}^E \int_0^L \psi_j (-\omega_i^2 \psi_i) dx_3,$$

$$= A C_{33}^E \omega_i^2 (\psi_i, \psi_j)_{L^2(D)} = \begin{cases} 0 & \text{if } i \neq j, \\ k_i := A C_{33}^E \omega_i^2 \int_0^L \psi_i^2(x_3) dx_3 & \text{if } i = j. \end{cases}$$

The set of ordinary differential equations that are satisfied by the time-varying coefficients $\{u_{3,j}\}_{j=1}^{n_d}$ are written in standard form by dividing the jth equation by the modal mass m_j. We see then that

$$\begin{bmatrix} 1 & 0 & \ldots & 0 \\ 0 & 1 & \ldots & \vdots \\ \vdots & \ldots & \ddots & 0 \\ 0 & \ldots & 0 & 1 \end{bmatrix} \begin{Bmatrix} \ddot{u}_{3,1}(t) \\ \vdots \\ \ddot{u}_{3,n_d}(t) \end{Bmatrix} + \begin{bmatrix} \omega_{n,1}^2 & 0 & \ldots & 0 \\ 0 & \omega_{n,2}^2 & \ldots & \vdots \\ \vdots & \ldots & \ddots & 0 \\ 0 & \ldots & 0 & \omega_{n,n_d}^2 \end{bmatrix} \begin{Bmatrix} u_{3,1}(t) \\ \vdots \\ u_{3,n_d}(t) \end{Bmatrix} = \begin{Bmatrix} 0 \\ \vdots \\ 0 \end{Bmatrix},$$

and $\omega_{n,j} := \sqrt{\frac{k_j}{m_j}}$ is the natural frequency of the jth modal degree of freedom $u_{3,j}$. It is well known [21] that the solution of the jth row of this matrix equation is given in closed form by

$$u_{3,j}(t) = \frac{\overline{v}_{0,j}}{\omega_{n,j}} \sin \omega_{n,j} t + \bar{u}_{0,j} \cos \omega_{n,j} t, \tag{8.22}$$

where $\omega_{n,j} := \sqrt{\frac{m_j}{k_j}} = \sqrt{\frac{C_{33}^E}{\rho_m}} \omega_j$ is the natural frequency of the jth degree of freedom, $\bar{u}_{0,j}$ is the initial condition of $u_{3,j}(t)$, and $\overline{v}_{0,j}$ is the initial condition of $\dot{u}_{3,j}(t)$. Since displacement is given by the sum $u_3(t, x_3) = \psi_j(x_3) u_{3,j}(t)$, the

(Continued)

Example 8.3.3 (Continued)

initial conditions satisfy the equations

$$u_3(0,x_3) = \psi_j(x_3)u_{3,j}(0) = \psi_j(x_3)\bar{u}_{0,j} = \bar{u}_0(x_3),$$
$$\dot{u}_3(0,x_3) = \psi_j(x_3)\dot{u}_{3,j}(0) = \psi_j(x_3)\overline{v}_{0,j} = \overline{v}_0(x_3).$$

We premultiply each of these equations by an arbitrary mode shape ψ_i, integrate over $[0, L]$, and use the orthogonality of the basis to solve for the initial conditions that hold for the degrees of freedom

$$\bar{u}_{0,j} = \frac{\int_0^L \psi_j(x_3)\bar{u}_0(x_3)dx_3}{\sqrt{\int_0^L \psi_j^2(x_3)dx_3}},$$
$$\overline{v}_{0,j} = \frac{\int_0^L \psi_j(x_3)\overline{v}_0(x_3)dx_3}{\sqrt{\int_0^L \psi_j^2(x_3)dx_3}}. \tag{8.23}$$

See Theorem S.1.1 in the Appendix for a justification of this representation. In summary, the homogeneous solution of the equations of motion for the axial linearly elastic rod is achieved by substituting the initial conditions in Equation 8.23 into the solution for each degree of freedom in Equation 8.22, and then substituting the result into the modal summation in Equation 8.21 to obtain $u_3(t, x_3)$:

$$u_3(t,x_3) = \left(\frac{\overline{v}_{0,j}}{\omega_{n,j}} \sin \omega_{n,j} t + \bar{u}_{0,j} \cos \omega_{n,j} t\right) \sin \frac{j\pi x_3}{L}.$$

In the next example we derive an approximation of the weak solution for the approximate governing equations of a linearly elastic, composite, Bernoulli–Euler beam using a basis of eigenfunctions.

Example 8.3.4 We can write the generalized eigenvalue problem in Equation 8.20 when the structural properties are uniform as

$$C_{11}^E I \frac{d^4\psi}{dx^4} + \lambda \rho_m A \psi = 0$$

subject to the boundary conditions

$$\psi(0) = 0, \qquad \frac{d^2\psi}{dx^2}(L) = 0,$$
$$\frac{d\psi}{dx}(0) = 0, \qquad \frac{d^3\psi}{dx^3}(L) = 0.$$

This generalized eigenvalue problem is rewritten as an ordinary eigenvalue problem

$$\frac{d^4\psi}{dx^4} = \omega^4\psi$$

with $\omega^4 := -\frac{\rho_m A}{C_{11}^E I}\lambda$. Example S.1.4 in the Appendix solves for the eigenfunctions $\psi_j(x)$ and eigenvalues ω_j for $j = 1, \ldots, \infty$, and the lengthy calculation will not be repeated here. The Galerkin approximation $u_3(t, x_1) = \psi_j(x_1)u_{3,j}(t)$, with the summation convention over $j = 1, \ldots, n_d$, of the weak governing equations then generates a mass matrix given by

$$\begin{aligned} m_{ij} &= \int_0^L \rho_m A N_i(x_1)N_j(x_1)dx_1 = \rho_m A \int_0^L \psi_i(x_1)\psi_j(x_1)dx_1 \\ &= \begin{cases} m_i := \rho_m A \int_0^L \psi_i^2(x_1)dx_1 & \text{if } i = j, \\ 0 & \text{otherwise.} \end{cases} \end{aligned}$$

This expression shows, similarly to Example 8.3.3, that the mass matrix is diagonal. The jth modal mass m_j is located at the diagonal element of the jth row. The stiffness matrix can also be shown to be diagonal by integrating by parts, employing the boundary conditions, and making use of the orthogonality of the modes ψ_j,

$$\begin{aligned} k_{ij} &= \int_0^L C_{11}^E I \frac{d^2N_i}{dx_1^2}\frac{d^2N_j}{dx_1^2}dx_1 = C_{11}^E I \int_0^L \frac{d^2\psi_i}{dx_1^2}\frac{d^2\psi_j}{dx_1^2}dx_1, \\ &= C_{11}^E I \left\{ \left.\frac{d^2\psi_i}{dx_1^2}\frac{d\psi_j}{dx_1}\right|_0^L - \left.\frac{d^3\psi_i}{dx_1^3}\psi_j\right|_0^L + \int_0^L \frac{d^4\psi_i}{dx_1^4}\psi_j dx_1 \right\}, \\ &= C_{11}^E \omega_i^4 \int_0^L \psi_i\psi_j dx_1 = \begin{cases} k_i := C_{11}^E I\omega_i^4 \int_0^L \psi_i^2 dx_1 & \text{if } i = j, \\ 0 & \text{otherwise.} \end{cases} \end{aligned}$$

As in Example 8.3.3, the jth modal stiffness k_j is located at the diagonal element of the jth row. The the approximation of the weak solution in terms of the DOF variables $u_{3,j}(t)$ can be rearranged into the diagonal system

$$\begin{bmatrix} 1 & 0 & \ldots & 0 \\ 0 & 1 & \ldots & 0 \\ \vdots & \vdots & \ddots & \vdots \\ 0 & 0 & \ldots & 1 \end{bmatrix} \begin{Bmatrix} \ddot{u}_{3,1}(t) \\ \vdots \\ \ddot{u}_{3,n_d}(t) \end{Bmatrix} + \begin{bmatrix} \omega_{n,1}^2 & 0 & \ldots & 0 \\ 0 & \omega_{n,2}^2 & \ldots & 0 \\ \vdots & \vdots & \ddots & \vdots \\ 0 & 0 & \ldots & \omega_{n,n_d}^2 \end{bmatrix} \begin{Bmatrix} u_{3,1}(t) \\ \vdots \\ u_{3,n_d}(t) \end{Bmatrix} = \begin{Bmatrix} 0 \\ \vdots \\ 0 \end{Bmatrix}.$$

(Continued)

Example 8.3.4 (Continued)

In these equations we define the natural frequency ω_{n_j} of the jth modal degree of freedom as

$$\omega_{n_j} := \sqrt{\frac{C_{11}^E I}{\rho_m A}\omega_j^2}. \tag{8.24}$$

As in Example 8.3.3, the analytical form of the homogeneous solution of the jth row of the above equations is given by

$$u_{3_j}(t) = \frac{\overline{v}_{0,j}}{\omega_{n_j}} \sin \omega_{n_j} t + \bar{u}_{0,j} \cos \omega_{n_j} t, \tag{8.25}$$

with the initial conditions determined from the identities

$$\begin{aligned}
\bar{u}_{0,j} &= \frac{\int_0^L \psi_j(x_1)\bar{u}_0(x_1)dx_1}{\sqrt{\int_0^L \psi_j^2(x_1)dx_1}}, \\
\overline{v}_{0,j} &= \frac{\int_0^L \psi_j(x_1)\overline{v}_0(x_1)dx_1}{\sqrt{\int_0^L \psi_j^2(x_1)dx_1}}.
\end{aligned} \tag{8.26}$$

The final homogeneous solution for the displacement $u_3(t, x_1)$ is achieved by substituting Equations 8.24 and 8.26 into Equation 8.25, and subsequently substituting 8.25 into the Galerkin approximation $u(t, x_1) = \psi_j(x_1)u_{3_j}(t)$ for $j = 1, \dots, n_d$, to obtain

$$\begin{aligned}
u_3(t, x_1) &= \left(\frac{\overline{v}_{0,j}}{\omega_{n_j}} \sin \omega_{n_j} t + \bar{u}_{0,j} \cos \omega_{n_j} t \right) \psi_j(x_1), \\
&= \left(\frac{\overline{v}_{0,j}}{\omega_{n_j}} \sin \omega_{n_j} t + \bar{u}_{0,j} \cos \omega_{n_j} t \right) \cdots \\
&\quad \times (\cosh \omega_j x_1 - \cos \omega_j x_1 - \alpha_j(\sinh \omega_j x_1 - \sin \omega_j x_1)).
\end{aligned}$$

The determination of ω_j and α_j that appear in this solution is summarized in Example S.1.4 in the Appendix.

Examples 8.3.3 and 8.3.4 consider the approximation of the free vibration of a linearly elastic rod and beam, respectively. Choosing the basis of the Galerkin approximation to be the eigenfunctions of the spatial operator proved to be advantageous. The degrees of freedom of the Galerkin approximation are shown to satisfy equations that are diagonal. In the next two examples, we consider the Galerkin approximation of linearly piezoelectric structures that are driven by the input voltage V.

Example 8.3.5 In this example we employ an analogous strategy to that in Example 8.3.3, but now consider a linearly piezoelectric rod that is driven by a input voltage in

$$\rho_m A \frac{\partial^2 u_3}{\partial t^2} = \frac{\partial}{\partial x_3}\left(AC_{33}^E \frac{\partial u_3}{\partial x_3}\right) + \frac{\partial}{\partial x_3}\left(\frac{Ae_{33}\chi_{[a,b]}}{b-a}\right)V,$$

subject to the boundary conditions in the table in Example 8.3.1, and subject to the initial conditions as described in Example 8.3.3. We choose the Galerkin approximation $u_3(t,x_3) = u_{3,j}(t)\psi_j(x_3)$ for $j = 1,\dots,n_d$ where the eigenpairs $\left(-\omega_j^2, \psi_j\right) = \left(-(\frac{j\pi}{L})^2, \sin\frac{j\pi x_3}{L}\right)$ solve the eigenvalue problem in Equations 8.50, 8.51, 8.52 for $j = 1,\dots\infty$. Since the eigenfunctions ψ_j for $j = 1,\dots,n_d$ satisfy the boundary conditions $\psi_j(0) = \psi_j(L) = 0$, the degrees of freedom $u_{3,j}(t)$ of the Equations 8.16 satisfy

$$\begin{bmatrix} 1 & 0 & \dots & 0 \\ 0 & 1 & \dots & \vdots \\ \vdots & \dots & \ddots & 0 \\ 0 & \dots & 0 & 1 \end{bmatrix} \begin{Bmatrix} \ddot{u}_{3,1}(t) \\ \vdots \\ \ddot{u}_{3,n_d}(t) \end{Bmatrix} + \begin{bmatrix} \omega_{n,1}^2 & 0 & \dots & 0 \\ 0 & \omega_{n,2}^2 & \dots & \vdots \\ \vdots & \dots & \ddots & 0 \\ 0 & \dots & 0 & \omega_{n,n_d}^2 \end{bmatrix} \begin{Bmatrix} u_{3,1}(t) \\ \vdots \\ u_{3,n_d}(t) \end{Bmatrix} = \begin{Bmatrix} \frac{b_1}{m_1} \\ \vdots \\ \frac{b_{n_d}}{m_{n_d}} \end{Bmatrix} V.$$

where $m_j := \rho_m A \int_0^L \psi_j^2(x_3)dx_3$, $k_j := AC_{33}^E\omega_j^2 \int_0^L \psi_j^2(x_3)dx_3$, $\omega_{n,j}^2 = \frac{m_j}{k_j} = \frac{C_{33}^E}{\rho_m}\omega_j^2$, and $b_j := -\int_0^L \frac{Ae_{33}\chi_{[a,b]}}{b-a}\psi_j'(x_3)dx_3$. These equations are, as in Example 8.3.4, uncoupled. The ith row is

$$\ddot{u}_{3,j}(t) + \omega_{n,j}^2 u_{3,j}(t) = \frac{b_j}{m_j}V(t). \tag{8.27}$$

The total solution for the jth degree of freedom $u_{3,j}(t)$ has the form

$$u_{3,j}(t) = \underbrace{\mathbb{C}_{1,j}\cos\omega_{n,j}t + \mathbb{C}_{2,j}\sin\omega_{n,j}t}_{\text{homogeneous}} + \underbrace{u_{3,j}^{\mathrm{p}}(t)}_{\text{particular}}, \tag{8.28}$$

where $u_{3,j}^{\mathrm{p}}$ is a particular solution that satisfies the inhomogeneous Equation 8.27, and the homogeneous solution satisfies Equation 8.27 when the right hand side is set to zero. The initial conditions on the degrees of freedom $\bar{u}_{0,j}$ and $\bar{v}_{0,j}$ are computed as shown in Equation 8.23. The set of coefficients $\mathbb{C}_{1,j}, \mathbb{C}_{2,j}$ can then be solved for $j = 1,\dots,n_d$ by enforcing the initial conditions $u_{3,j}(0) = \bar{u}_{0,j}$ and $\dot{u}_{3,j}(0) = \bar{v}_{0,j}$ in Equation 8.27. It is important to note that the total solution $u_{3,j}(t)$ in Equation 8.28 will then depend on both the voltage V applied on the right hand side of Equation 8.27 and on the initial condition functions $\bar{u}_0$ and $\bar{v}_0$.

(Continued)

Example 8.3.5 (Continued)

Let us consider now a special case where we seek to find the steady state response to a harmonic input, either $V(t) = V_0 \cos \varpi t$ or $V(t) = V_0 \sin \varpi t$ where ϖ is the driving frequency and V_0 is the magnitude of the input voltage. This calculation is carried out most frequently in engineering vibrations using complex frequency response functions (FRFs). See Sections S.1.1 and S.1.2 in the Appendix for a presentation of the theoretical foundations of FRFs. We suppose, in the interest of generating a clear and concise example, that the initial condition functions $\bar{u}_0$ and $\overline{v}_0$ are chosen as $u_3(0, x_3) = \psi_j(x_3) u^{\mathrm{p}}_{3,j}(0) = \bar{u}_0(x_3)$ and $\dot{u}(0, x_3) = \psi_j(x_3) \dot{u}^{\mathrm{p}}_{3,j}(0) = \overline{v}_0(x_3)$ for all $x_3 \in \Omega = [0, L]$. In this case the constants $\mathbb{C}_{1j} = \mathbb{C}_{2j} = 0$ for $j = 1, \ldots, n_d$. We will use the complex response method [11] to find the steady state response. Instead of just Equation 8.27, we simultaneously consider the two equations

$$\ddot{u}_{3,j}(t) + \omega^2_{n,j} u_{3,j}(t) = \frac{b_j}{m_j} V_0 \cos \varpi t, \tag{8.29}$$

$$\ddot{v}_{3,j}(t) + \omega^2_{n,j} v_{3,j}(t) = \frac{b_j}{m_j} V_0 \sin \varpi t. \tag{8.30}$$

The responses $u_{3,j}(t)$ and $v_{3,j}(t)$ are real-valued functions of time in this pair of equations. If we multiply the second equation by $\hat{j} = \sqrt{-1}$ and add the resulting two equations, we generate the complex response equation

$$\ddot{z}_{3,j}(t) + \omega^2_{n,j} z_{3,j}(t) = \frac{b_j}{m_j} V_0 e^{\hat{j}\varpi t} \tag{8.31}$$

in terms of the complex state variable $z_{3,j}(t) = u_{3,j}(t) + \hat{j} v_{3,j}(t)$. If we solve for the complex response $z_{3,j}(t)$ in Equation 8.31, we can can set $u_{3,j}(t) = \mathrm{re}(z_{3,j}(t))$ to get the solution due to the harmonic input $V_0 \cos \Omega t$ in Equation 8.29. On the other hand, if we want the to get the solution due to the input $V_0 \sin \varpi t$ in Equation 8.30, we set $u_{3,j}(t) = \mathrm{im}(z_{3,j}(t))$. We construct the solution of the complex response $z_{3,j}$ of DOF j by assuming that it has the form

$$z_{3,j}(t) = \mathbb{Z}_{3,j} e^{\hat{j}\varpi t}. \tag{8.32}$$

for some complex constant $\mathbb{Z}_{3,j}$. When we substitute Equation 8.32 into the complex response Equation 8.31, it follows that

$$\mathbb{Z}_{3j} = \frac{1}{(\omega^2_{n,j} - \varpi^2)} \frac{b_j}{m_j} V_0 = \frac{1}{|\omega^2_{n,j} - \varpi^2|} e^{\hat{j}\beta_j} \frac{b_j}{m_j} V_0$$

with $\beta_j := \arg((\omega_{n,j}^2 - \varpi^2))$. In other words, the phase $\beta_j = 0$ or $\beta_j = \pi$, depending on whether the real-valued constant $(\omega_{n,j}^2 - \varpi^2)$ is positive or negative, respectively. We simplify notation a bit by introducing the same symbol $u_{3,j}(t) := z_{3,j}(t)$ for the complex modal response and real modal response of mode j. It is the complex response of the jth mode that is due to the complex harmonic input $V(t) = V_0 e^{\hat{j}\varpi t}$. By definition then, we have the complex modal response given by

$$u_{3,j}(t) = \frac{1}{|\omega_{n,j}^2 - \varpi^2|} e^{\hat{j}\beta_j} \frac{b_j}{m_j} V_0 e^{\hat{j}\varpi t}.$$

If we want the real modal response $u_{3,j}(t)$ due to a sinusoidal input $V(t) = V_0 \sin(\Omega t)$ in Equation 8.30, then it is given by

$$\begin{aligned}
u_{3,j}(t) &= \text{im}(z_{3,j}(t)) = \text{im}\{\mathbb{Z}_{3,j} e^{\hat{j}\varpi t}\}, \\
&= \text{im}\left(\frac{1}{|\omega_{n,j}^2 - \varpi^2|} e^{\hat{j}\beta_j} \frac{b_j}{m_j} V_0 e^{\hat{j}\varpi t} \right) \\
&= \frac{1}{|\omega_{n,j}^2 - \varpi^2|} \frac{b_j}{m_j} V_0 \sin(\varpi t + \beta_j).
\end{aligned}$$

On the other hand, if we want the real modal response $u_{3,j}(t)$ due to the the voltage input $V(t) = V_0 \cos(\varpi t)$ in Equation 8.29, we find that

$$u_{3,j}(t) = \frac{1}{|\omega_{n,j}^2 - \varpi^2|} \frac{b_j}{m_j} V_0 \cos(\varpi t + \beta_j).$$

We conclude that, in the former case when we are interested in the sinusoidal input, the steady state response $u_3(t, x_3)$ for any point $x_3 \in [0, L]$ along the rod is given by the sum over $j = 1, \ldots, n_d$ in

$$u(t, x_3) = \frac{1}{|\omega_{n,j}^2 - \varpi^2|} \frac{b_j}{m_j} V_0 \sin(\varpi t + \beta_j) \sin \frac{j\pi x_3}{L}.$$

In the latter case, when we seek to find the real steady state response $u(t, x_3)$ to the input $V(t) := V_0 \sin(\varpi t)$, the response is then given by the sum over $j = 1, \ldots, n_d$ in

$$u(t, x_3) = \frac{1}{|\omega_{n,j}^2 - \varpi^2|} \frac{b_j}{m_j} V_0 \cos(\varpi t + \beta_j) \sin \frac{j\pi x_3}{L}.$$

Example 8.3.6 In this example we study the response of the piezoelectric composite Bernoulli–Euler beam that is driven by the input voltage $V(t)$ and satisfies the strong form of the governing Equations 8.2 or 7.23

$$\rho_m A \frac{\partial^2 u_3}{\partial t^2} = -\frac{\partial^2}{\partial x_1^2}\left(C_{11}^E I \frac{\partial^2 u_3}{\partial x_1^2} \right) + \frac{\partial^2}{\partial x_1^2}\left(\frac{\kappa e_{31} \chi_{[a,b]}}{t_p} \right) V \tag{8.33}$$

subject to the boundary conditions

$$\begin{aligned} u_3(t,0) &= 0, & \mathcal{S}(t,L) &= 0,\\ \frac{\partial u_3}{\partial t}(t,0) &= 0, & \mathcal{M}(t,L) &= 0, \end{aligned}$$

for $t \in \mathbb{R}^+$ and to the initial conditions

$$\begin{aligned} u_3(0,x_3) &= \bar{u}_0(x_3),\\ \frac{\partial u_3}{\partial t}(0,x_3) &= \overline{v}_0(x_3), \end{aligned}$$

for all $x_3 \in D \equiv [0,L]$. The Galerkin method uses eigenfunctions ψ_i from Example 8.3.4 as the basis of an approximation that takes the form

$$u_3(t,x_1) = \psi_j(x_1) u_{3,j}(t)$$

with $t \in \mathbb{R}^+$ and the summation is over $j = 1, \dots, n_d$. When all of the boundary conditions are enforced, the approximation of the weak form of the governing equations can be rearranged in the diagonal form

$$\begin{bmatrix} 1 & & \\ & \ddots & \\ & & 1 \end{bmatrix} \begin{Bmatrix} \ddot{u}_{3,1}(t) \\ \vdots \\ \ddot{u}_{3,n_d}(t) \end{Bmatrix} + \begin{bmatrix} \omega_{n,1}^2 & & \\ & \ddots & \\ & & \omega_{n,n_d}^2 \end{bmatrix} \begin{Bmatrix} u_{3,1}(t) \\ \vdots \\ u_{3,n_d}(t) \end{Bmatrix} = \begin{Bmatrix} \dfrac{b_1}{m_1} \\ \vdots \\ \dfrac{b_{n_d}}{m_{n_d}} \end{Bmatrix} V(t). \tag{8.34}$$

In this equation the modal mass $m_j = \rho_m A \int_0^L \psi_j^2 dx_1$, the modal stiffness $k_j = C_{11}^E I \omega_j^4 \int_0^L \psi_j^2 dx_1$, the control influence $b_j = \int_0^L \frac{\kappa e_{31} \chi_{[a,b]}}{t_p} \frac{d^2 \psi_j}{dx_1^2} dx_1$, and $\omega_{n,j}^2 = \frac{k_j}{m_j} = \frac{C_{11}^E I}{\rho_m A} \omega_j^4$. As in Equation 8.27, the jth row of Equations 8.34 has the total solution

$$u_{3,j}(t) = \underbrace{(\mathbb{C}_{1,j} \cos \omega_{n,j} t + \mathbb{C}_{2,j} \sin \omega_{n,j} t)}_{\text{homogeneous}} + \underbrace{u_{3,j}^{\mathrm{p}}(t)}_{\text{particular}}.$$

In this equation $u_{3,j}^{\mathrm{h}}$ is the homogeneous solution that satisfies Equation 8.34 when the right hand sight is set to zero, and $u_{3,j}^{\mathrm{p}}$ is the particular solution obtained by satisfying Equation 8.34 including the nonzero right hand side.

As in Example 8.3.5, the coefficients $\mathbb{C}_{1,j}$ and $\mathbb{C}_{2,j}$ can be determined from the initial conditions on the variables $u_{3,j}(t)$, $\frac{du_{3,j}}{dt}(t)$ and in terms of the particular solution $u_j^{\mathrm{p}}(t)$. The general form of the approximate solution then is given by the summation

$$u(t,x_1) = (\mathbb{C}_{1,j}\cos\omega_{n,j}t + \mathbb{C}_{2,j}\sin\omega_{n,j}t + u_{3,j}^{\mathrm{p}}(t))\psi_j(x_1)$$

with the summation convention over $j = 1, \ldots, n_d$

Let us again find the form of the steady state response when all the coefficients $\mathbb{C}_{1,j}, \mathbb{C}_{2,j} \equiv 0$, as carried out in Example 8.3.3 for the composite piezoelectric rod. When we follow the steps analogous to those in Example 8.3.3, we find that the complex response equation is given by the summation over $j = 1, \ldots, n_d$ in

$$u(t,x_1) = \frac{1}{|\omega_{n,j}^2 - \varpi^2|} e^{\hat{j}\beta_j} \frac{b_j}{m_j} V_0 e^{\hat{j}\varpi t}\psi_j(x_1),$$

and each phase angle β_j given by

$$\beta_j = \arg(\omega_{n,j}^2 - \varpi^2).$$

It follows that the real steady state response to the harmonic input $V(t) = V_0 \sin \varpi t$ is given by

$$u(t,x_1) = \frac{1}{|\omega_{n,j}^2 - \varpi^2|} \frac{b_j}{m_j} V_0 \sin(\varpi t + \beta_j)\psi_j(x_1),$$

with the summation convention over $j = 1, \ldots, n_d$. The steady state solution to the harmonic input $V_0 \cos(\varpi t)$ is analogous.

Example 8.3.7 In this example we assume that energy dissipation is governed by a viscous damping model. Specifically, we assume that the damping matrix c_{ij} constructed from the bilinear form $a_1(\cdot,\cdot)$ in Equation 8.9 is diagonalized by the mode shapes ψ_i in the sense that

$$c_{ij} = \begin{cases} c_i := a_1(\psi_i, \psi_i) & \text{if} \quad i = j, \\ 0 & \text{if} \quad i \neq j, \end{cases}$$

where the modal damping coefficient c_j is given for $j = 1, \ldots, n_d$. The analysis that follows can be applied to either the axial composite in Example 8.3.5 or the composite beam in Example 8.3.6. Only the values of the constants $m_i, c_i, b_i, \omega_{n,i}, \omega_i^2$ and eigenfunctions ψ_i, which can be found in

(Continued)

Example 8.3.7 (Continued)

the corresponding Examples 8.3.5 and 8.3.6, differ in the two cases. We let the generic spatial variable x in this example denote either x_3 or x_1 when considering the either the axial or beam composite, respectively. In either case, the equations that are satisfied by the degrees of freedom $u_{3,j}$ in the Galerkin approximation $u_{3,j}(t,x) = \psi_j(x)u_{3,j}(t)$ are written in the form

$$\begin{bmatrix} 1 & & \\ & \ddots & \\ & & 1 \end{bmatrix}\begin{Bmatrix} \ddot{u}_1(t) \\ \vdots \\ \ddot{u}_{n_d}(t) \end{Bmatrix} + \begin{bmatrix} 2\xi_1\omega_{n,1} & & \\ & \ddots & \\ & & 2\xi_{n_d}\omega_{n,n_d} \end{bmatrix}\begin{Bmatrix} \dot{u}_1(t) \\ \vdots \\ \dot{u}_{n_d}(t) \end{Bmatrix}$$

$$+\begin{bmatrix} \omega_{n,1}^2 & & \\ & \ddots & \\ & & \omega_{n,n_d}^2 \end{bmatrix}\begin{Bmatrix} u_1(t) \\ \vdots \\ u_{n_d}(t) \end{Bmatrix} = \begin{Bmatrix} \dfrac{b_1}{m_1} \\ \vdots \\ \dfrac{b_{n_d}}{m_{n_d}} \end{Bmatrix} V(t).$$

where ξ_j is the damping ratio of the jth degree of freedom, $2\xi_j\omega_{n,j} = \frac{c_j}{m_j}$, the jth natural frequency $\omega_{n,j} = \sqrt{\frac{k_j}{m_j}}$, m_j is the jth modal mass, k_j is the jth modal stiffness, and b_j is the jth entry in the control influence operator. Suppose that we again determine the steady state response due to a harmonic input as in Example 8.3.5 or 8.3.6. The complex response $z_{3,j}(t) = u_{3,j}(t) + \hat{j}v_{3,j}(t)$ of the jth degree of freedom satisfies the equation

$$\ddot{z}_{3,j}(t) + 2\xi_j\omega_{n,j}\dot{z}_{3,j}(t) + \omega_{n,j}^2 z_{3,j}(t) = \frac{b_j}{m_j}V_0 e^{\hat{j}\varpi t}$$

where ϖ is the driving frequency of the complex voltage $V(t) := V_0 e^{\hat{j}\varpi t}$ and V_0 is the amplitude of the voltage.

When we substitute the assumed form of the complex response $z_{3,j}(t) := \mathbb{Z}_{3,j}e^{\hat{j}\varpi t}$ into the governing equations, we see that

$$\mathbb{Z}_{3,j} = \frac{1}{\sqrt{(\omega_{n,j}^2 - \varpi^2)^2 + (2\xi_j\omega_{n,j}\varpi)^2}} e^{\hat{j}\beta_j}\frac{b_j}{m_j}V_0,$$

$$\beta_j = -\arctan\frac{2\xi_j\omega_{n,j}\varpi}{(\omega_{n,j}^2 - \varpi^2)}.$$

The complex response of mode j is then defined as $u_{3,j}(t) := z_{3,j}(t)$ in the equation

$$u_{3,j}(t) = \frac{1}{\sqrt{(\omega_{n,j}^2 - \varpi^2)^2 + (2\xi_j \omega_{n,j} \varpi)^2}} e^{\hat{j}\beta_j} \frac{b_j}{m_j} V_0 e^{\hat{j}\varpi t}.$$

The complex response $u_3(t, L)$ at the tip due to the real input $V(t) = V_0 e^{\hat{j}\varpi t}$ is computed to be

$$u_3(t, L) = \underbrace{\frac{1}{\sqrt{(\omega_{n,j}^2 - \varpi^2)^2 + (2\xi_j \omega_{n,j} \varpi)^2}} e^{\beta_j} \frac{b_j}{m_j} \psi_j(L)}_{H(\hat{j}\varpi)} V_0 e^{\hat{j}\varpi t},$$

with the summation convention over $j = 1, \dots, n_d$. The function $H(\hat{j}\Omega)$ is the complex frequency response function from the harmonic input having amplitude V_0 to the harmonic response at the end $x = L$ of the structure. Similar expressions can be derived for any station $x \in [0, L]$ along the structure. As before, the steady state response can also be written in the form

$$u_3(t, L) = \frac{1}{\sqrt{(1 - r_j^2)^2 + (2\xi_j r_j)^2}} e^{\hat{j}\beta_j} \frac{b_j}{k_j} \psi_j(L) V_0 e^{\hat{j}\varpi t}$$

with $r_j := \varpi / \omega_{n,j}$ and the summation convention over $j = 1, \dots, n_d$.

8.3.2 Finite Element Approximations

We have seen in Section 8.3.1 that for some active composites that exhibit simple geometry and uniform material properties, it can be advantageous to utilize eigenfunctions to construct approximate solutions. If the modal approach is feasible, the governing equations are diagonal, which enables analytic solution of the approximate governing equations. For more general problems, such as those in which the domain is irregular in two or three dimensional continuum piezoelectricity, or in when the material properties are not uniform, modal approximation methods can be more difficult, if not impossible, to employ. In this section we discuss finite element methods that provide a flexible alternative to modal approximations. These techniques are well suited to more general geometries or nonuniform material properties.

As emphasized in Section 8.3.1, one of the crucial advantages of the modal approach is that the eigenfunction bases have biorthogonality properties: they render the mass and stiffness matrices diagonal. In some sense this is an ideal situation when seeking to find an analytic solution. On the other hand, if other types of basis functions that are supported over the entire domain are used to build approximations, then the mass and stiffness matrices are typically full. That is, in such a general case, $m_{ij} \neq 0, k_{ij} \neq 0$ for $i, j = 1, \ldots, n_d$. This fact can be shown directly from the definition $m_{ij} := \int_0^L \rho_m A N_i(x) N_j(x) dx$. If the support of the basis functions N_i and N_j overlap, then the integral is not guaranteed in general to be equal to zero.

Finite element methods introduce bases that have been derived to provide a systematic approximation technique that emulates, in a rough sense, some of the best properties of orthogonal basis functions. Each finite element basis function N_i for $i = 1, \ldots, n_d$ is supported on some small set. We say that the basis functions N_i have compact support. So, although the basis functions N_i and N_j may not be orthogonal for all i, j, whenever the i, j correspond to basis functions that do not overlap, we have $m_{ij} = k_{ij} = 0$. The result is that the matrices m_{ij} and k_{ij} are banded and sparse. In addition, the finite element method has been under development and study for many decades, and large collections of basis functions have been derived that have various approximation properties and compact support. Approximation properties are usually established by showing that a family of finite elements contain piecewise polynomials or splines of a certain order. We will review only two classes of finite elements in this text, those that are convenient and popular for the axial rod and beam models. The reader is referred to [11] or [6] for a good introduction to the finite element method. A more detailed account, including a discussion of the more commonly available finite element families, can be found in the well known textbooks [51] and [19].

We discuss the construction of finite element functions for Galerkin approximations in the next two examples.

Example 8.3.8 When we carefully inspect the form of the mass and stiffness matrices in Equation 8.11, we can infer some of the common properties that the basis functions N_j for $j = 1, \ldots, n_d$ must satisfy. We recall from Example 8.3.1 that N_i and $\frac{dN_i}{dx_3}$ must each be square integrable so that the integrals that appear in the mass matrix $m_{ij} := \int_0^L \rho_m A N_i N_j dx_3$ and the stiffness matrix $k_{ij} := \int_0^L A C_{33}^E N_i' N_j' dx_3$ exist. In particular, this is true if the functions N_j for $j = 1, \ldots, n_d$ are continuous on $[0, L]$ and piecewise differentiable. Figure 8.1 depicts a collection of C^0 piecewise linear finite elements, perhaps the simplest such finite elements that can be used as shape functions in this problem.

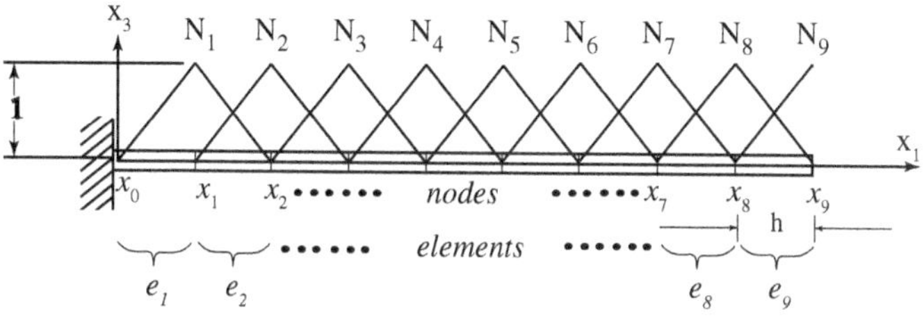

Figure 8.1 Piecewise linear C^0 finite elements

Note that the finite element shape functions have been selected so that the Galerkin approximation satisfies the geometric boundary condition $u_3(t,0) = N_i(0)u_{3,i}(t) = 0$. This definition is appropriate for an axial composite with fixed-free boundary conditions. The modification of the definitions of the finite elements at the ends of the bar for different boundary conditions such as fixed-fixed, free-free, or free-fixed, is straightforward. The polynomial form of the finite element shape functions makes it a simple matter to calculate the entries of the mass, stiffness, and control influence matrices m_{ij}, k_{ij}, and b_i. We partition the domain $[0, L]$ into elements $e_\ell := [x_{3,\ell-1}, x_{3,\ell}]$ and define the mesh width $h_\ell := x_{3,\ell} - x_{3,\ell-1}$ for $\ell = 1, \dots, n_d$, where $\{x_{3,\ell}\}_{\ell=0,\dots,n_d}$ are the nodes. Each shape function N_i is referred to as a hat function and is centered at node x_i. Entries of the mass, stiffness, and control influence matrices m_{ij}, k_{ij}, b_i are generated by summing contributions of the elemental mass, stiffness, and control influence matrices $m_{ij}^{e_\ell}, k_{ij}^{e_\ell}, b_i^{e_\ell}$ that are due to integration over each element e_ℓ. That is, we have

$$m_{ij} = \int_0^L \rho_m A N_i N_j dx_3 = \sum_{\ell=1}^{n_d} \int_{x_{\ell-1}}^{x_\ell} \rho_m A N_i N_j dx_3 = \sum_{\ell=1}^{n_d} m_{ij}^{e_\ell}$$

for the global mass matrix m_{ij}. Similar results hold for the global stiffness matrix k_{ij} and control matrix b_j. The elemental mass, stiffness, and control influence matrices can be computed directly and shown to be

$$m_{ij}^{e_\ell} = \begin{bmatrix} \int_{e_\ell} \rho_m A N_{\ell-1} N_{\ell-1} dx_3 & \int_{e_\ell} \rho_m A N_{\ell-1} N_\ell dx_3 \\ \int_{e_\ell} \rho_m A N_\ell N_{\ell-1} dx_3 & \int_{e_\ell} \rho_m A N_\ell N_\ell dx_3 \end{bmatrix} = \frac{\rho_m A h_\ell}{6} \begin{bmatrix} 2 & 1 \\ 1 & 2 \end{bmatrix},$$

$$k_{ij}^{e_\ell} = \begin{bmatrix} \int_{e_\ell} A C_{33}^E \frac{\partial N_{\ell-1}}{\partial x_3} \frac{\partial N_{\ell-1}}{\partial x_3} dx_3 & \int_{e_\ell} A C_{33}^E \frac{\partial N_{\ell-1}}{\partial x_3} \frac{\partial N_\ell}{\partial x_3} dx_3 \\ \int_{e_\ell} A C_{33}^E \frac{\partial N_\ell}{\partial x_3} \frac{\partial N_{\ell-1}}{\partial x_3} dx_3 & \int_{e_\ell} A C_{33}^E \frac{\partial N_\ell}{\partial x_3} \frac{\partial N_\ell}{\partial x_3} dx_3 \end{bmatrix}$$

(Continued)

Example 8.3.8 (Continued)

$$= \frac{AC_{33}^E}{h_\ell}\begin{bmatrix} 1 & -1 \\ -1 & 1 \end{bmatrix},$$

$$b_i^{e_\ell} = \begin{bmatrix} -\int_{e_\ell} \frac{Ae_{33}\chi_{[a,b]}}{b-a}\frac{\partial N_{\ell-1}}{\partial x_3}dx_3 \\ -\int_{e_\ell} \frac{Ae_{33}\chi_{[a,b]}}{b-a}\frac{\partial N_{\ell}}{\partial x_3}dx_3 \end{bmatrix} = \frac{Ae_{33}}{b-a}\begin{bmatrix} 1 \\ -1 \end{bmatrix},$$

respectively, for the choice of the linear finite elements. The elemental mass and stiffness matrices can be found in structural dynamics textbooks [11]. The calculation of the entries of the control influence matrix can be computed directly in closed form. A detailed discussion of the assembly process, whereby the entries of the elemental matrices $m_{ij}^{e_\ell}, k_{ij}^{e_\ell}, b_i^{e_\ell}$ are mapped and summed into the entries of the global matrices m_{ij}, k_{ij}, b_i can be found in textbooks on finite element methods such as [34] or [39].

When the assembly process is complete, the final form of the equations governing the degrees of freedom $u_{3,j}(t)$ in $u_3(t, x_3) = N_j(x_3)u_{3,j}(t)$ is given by

$$m_{ij}\ddot{u}_{3,j}(t) + c_{ij}\dot{u}_{3,j}(t) + k_{ij}u_{3,j}(t) = b_i V(t),$$

with the summation over $j = 1, \dots, n_d$ for each $i = 1, \dots, n_d$. The figures below depict results when the piezoelectric rod composite has the following system properties:

Table 8.1 Axial composite piezoelectric actuator, system properties.

Parameter	Description	Value	Units
C_{33}^E	Young's modulus	$6.7E10$	N/m^2
$d_{33} = e_{33}C_{33}^E$	Piezoelectric constant	$400E-12$	C/N
ρ_m	Density	7800	kg/m^2
m_t	Tip mass	0	kg
A	Cross sectional area	$1E-6$	m^2
L	Length	0.1	m
ξ_k for $k = 1, \dots, n_d$	Modal damping factor	0.001	Unitless

In Figure 8.2, the numerically computed natural frequencies $\omega_{n,k}^{n_d}$ are compared to their corresponding analytic values $\omega_{n,k}^a$ for $k = 1, \dots, 10$ when the Galerkin approximation includes $n_d = 10, 15, 20, 50, 100$ degrees of freedom. Several observations should be made from this figure. The numerical estimates converge from above to the analytic values. That is, for a given mode number k, the estimates of $\omega^a_{n,k}$ converge from $\omega^{10}_{n,k}$ that is associated with lowest dimension $n_d = 10$ to $\omega^{100}_{n,k}$ that is associated with the highest dimension $n_d = 100$. In addition, the numerical estimates $\omega^{n_d}_{n,k}$ of the analytic values $\omega^a_{n,k}$ of the natural frequencies are most accurate at low frequencies, and the error increases monotonically as the natural frequency $\omega^a_{n,k}$ increases. These observations are well documented in texts on vibrations or structural dynamics such as [21] or [11]. There is no difference in consideration of linearly piezoelectric electromechanical models since the eigenvalue problems are the same for linearly elastic or piezoelectric structures.

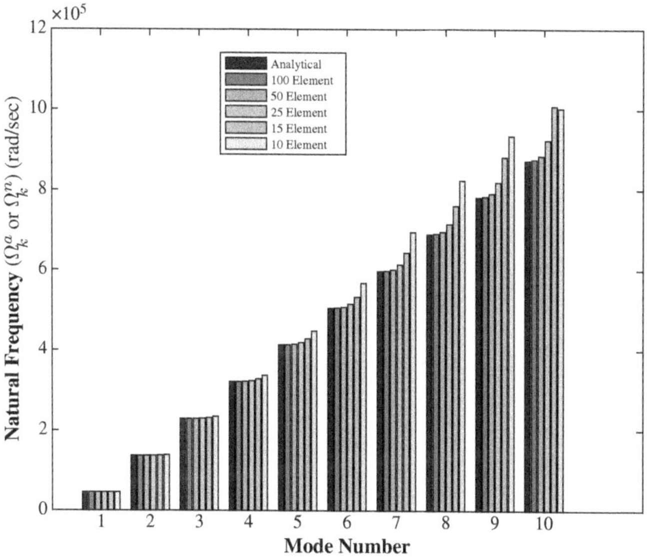

Figure 8.2 Comparison of analytical natural frequencies and numerical estimates of natural frequencies for $n_d = 10, 15, 25, 50, 100$

Figure 8.3 depicts the transient response of the axial piezoelectric rod with fixed-free boundary conditions when the voltage input is $V(t) := V_0 \cos(\varpi t)$ with $V_0 = 400$ volts and $\varpi = 10$ rad/sec. After a brief transient regime shown in both (a) and (b) below, the response quickly converges to a steady state oscillation. For each fixed location $x_3 \in [0, L]$ along the rod, the response $u_3(t, x_3)$ converges to a harmonic response proportional to $\cos(\varpi t + \beta)$ that has the same frequency as the driving frequency ϖ. This general feature of the

(Continued)

Example 8.3.8 (Continued)

response of a linear system to a harmonic input is well known and documented in standard texts such as [21] or [11]. The tip displacement $u_3(t, L)$ in (b) below is the trace of the surface plot in (a) at the tip $x_3 = L$.

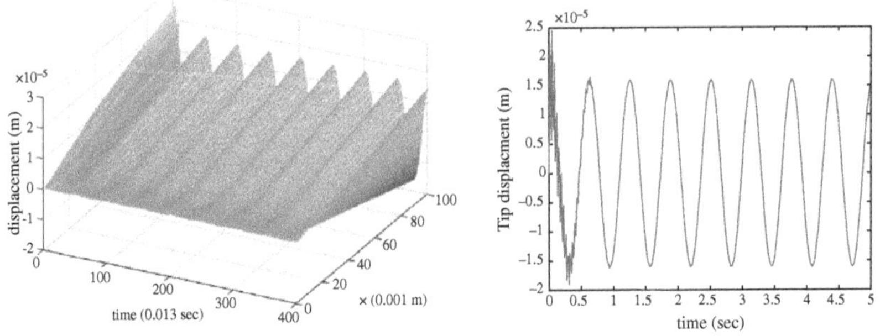

(a) Displacement $u_3(t, x_3)$ (b) Tip displacement $u_3(t, L)$

Figure 8.3 Transient response of axial piezoelectric specimen

While the plots above are generated via numerical simulation of the equations generated by a Galerkin approximation, it is possible to find closed form expressions for the response using the generalized algebraic eigenproblem summarized in Appendix S.1.3. Recall from Appendix S.1.3 that the generalized eigenvalues and eigenvectors $\{(\lambda_k, \psi_{j,k})_{j=1}^{n_d}\}_{k=1}^{n_d}$ satisfy the generalized algebraic eigenvalue problem

$$k_{ij}\psi_{j,k} = \lambda_k m_{ij}\psi_{j,k},$$

with the summation convention only over the indices $j = 1, \ldots, n_d$. The finite dimensional eigenvectors $\psi_{j,k}$, not to be confused with the infinite dimensional eigenfunctions $\psi_k(\cdot) \in L^2(\Omega)$ discussed in earlier examples, satisfy the mass and stiffness biorthogonality conditions

$$\psi_{i,\ell} m_{ij} \psi_{j,m} = m_\ell \delta_{\ell,m},$$
$$\psi_{i,\ell} k_{ij} \psi_{j,m} = k_\ell \delta_{\ell,m}.$$

We also assume that the damping matrix c_{ij} is diagonalized by the eigenvectors $\psi_{i,\ell}$ so that

$$\psi_{i,\ell} c_{ij} \psi_{j,m} = c_\ell \delta_{\ell,m}.$$

With these orthogonality conditions, we define the natural frequencies $\omega_{n,\ell}^2 = k_\ell / m_\ell$ and the damping ratios via $2\xi_\ell \omega_{n,\ell} = c_\ell / m_\ell$. When we express the degrees of freedom $u_{3,j}(t)$ for $j = 1, \ldots, n_d$ in terms of modal degrees of freedom η_ℓ for $\ell = 1, \ldots, n_d$ via the identity

$$u_{3,j}(t) = \psi_{j,\ell}\eta_\ell(t),$$

the orthogonality properties of the generalized eigenvectors yield the decoupled equations

$$m_\ell \ddot{\eta}_\ell(t) + c_\ell \dot{\eta}_\ell(t) + k_\ell \eta_\ell(t) = \psi_{j,\ell} b_j V(t)$$

with the summation convention only over $j = 1, \ldots, n_d$. The complex response of the ℓth modal coordinate η_ℓ to a harmonic input $V(t) = V_0 e^{\hat{j}\varpi t}$ can be written as

$$\eta_\ell(t) = H_{\ell j}(\hat{j}\varpi) b_j V_0 e^{\hat{j}(\varpi t)},$$

where $H_{\ell j}(\hat{j}\varpi)$ is given by

$$H_{\ell j}(\hat{j}\varpi) = \frac{\psi_{j,\ell}}{(1 - r_\ell^2) + 2\xi_\ell r_\ell \hat{j}} \frac{1}{k_\ell}.$$

In these equations r_ℓ is the ratio of the driving frequency ϖ to the ℓth natural frequency $\omega_{n,\ell}$, so that $r_\ell := \frac{\varpi}{\varpi_\ell}$. The mapping from a harmonic voltage input $V(t) := V_0 e^{\hat{j}\varpi t}$ to the harmonic output displacement $u_3(t, L)$ of the tip is given by the complex response

$$\begin{aligned} u_3(t, L) &= N_i(L)\psi_{i,\ell} H_{\ell j}(\hat{j}\varpi) b_j V_0 e^{\hat{j}\varpi t}, \\ &= N_i(L)\mathbb{H}_{i,j}(\hat{j}\varpi) b_j V_0 e^{\hat{j}\varpi t}, \end{aligned}$$

with the summation over $i, j, \ell = 1, \ldots, n_d$, and

$$\mathbb{H}_{i,j}(\hat{j}\varpi) := \frac{\psi_{i,\ell}\psi_{j,\ell}}{(1 - r_\ell^2) + 2\xi_\ell r_\ell \hat{j}} \frac{1}{k_\ell}.$$

with the summation convention over $\ell = 1, \ldots, n_d$. Finally, we define the frequency response function $H(\hat{j}\varpi) := N_i(L)\mathbb{H}_{i,j}(\hat{j}\varpi) b_j$ that maps the harmonic input $V(t) = V_0 e^{\hat{j}\varpi t}$ into the harmonic output at the tip of the beam $u_3(t, L)$. The Bode plot consists of plots of the magnitude plot of 20 log $|H(\hat{j}\varpi)|$ in decibels (db) and the plot of the phase $\arg(H(\hat{j}\varpi))$. In Figure 8.4 we compare the Bode plots of magnitude and phase for the lowest dimensional model with $n_d = 10$ and the highest dimensional model with $n_d = 100$.

Since the damping factor is $\xi_k = .001$ for all modes $k = 1, \ldots, n_d$, the piezoelectric structure is very lightly damped in this example. In practice, the damping factor for each mode can be obtained experimentally by

(Continued)

Example 8.3.8 (Continued)

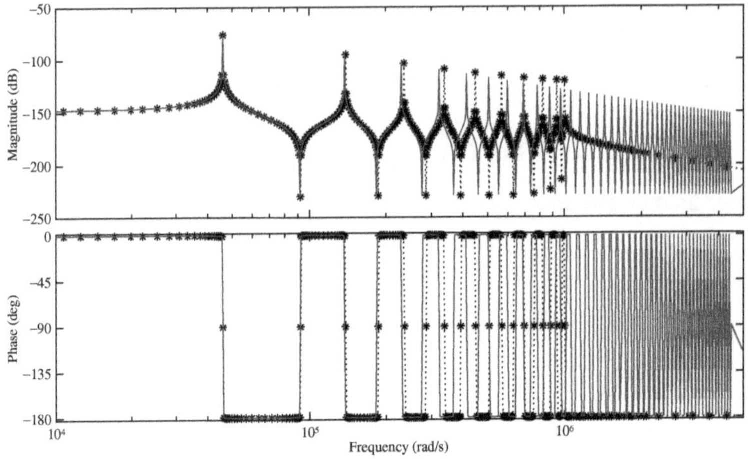

Figure 8.4 Comparison of Bode plots for models where the number of degrees of freedom $n_d = 10, 100$

the half-amplitude method or the logarithmic decrement method [11, 21], or by multiple degree of freedom modal identification techniques in the time or frequency domain. Figure 8.4 depicts the Bode plot for the model with $n_d = 10$ in black with '*' markers, while the corresponding Bode plot for $n_d = 100$ is a solid blue line without markers. Given our observations concerning the bar chart comparison of $\omega^{n_d}_{n,j}$ and $\omega^{a}_{n,j}$, it is not surprising that the Bode plots for $n_d = 10$ and $n_d = 100$ agree most closely at low frequencies. In fact, the Bode plots associated with $n_d = 10$ and $n_d = 100$ are indistinguishable over the range of driving frequency ϖ around the first three or so modes. The difference between the two models increases as the driving frequency ϖ grows.

Example 8.3.9 In this example we use finite element shape functions as the basis in a Galerkin approximation of the piezoelectric composite beam. The weak form of the Equations 8.11 have been studied in some detail in Example 8.3.2. The mass, stiffness, and control influence matrices are defined in 8.17, 8.18, and 8.19, respectively. We see that if the shape functions N_i and their second derivatives $\frac{dN_i^2}{dx_1^2}$ for $i = 1, \ldots, n_d$ are square integrable, then the integrals that define these matrices exist. We construct the Galerkin approximation

$$u_3 = u_3(t, x_1) = N_j(x_1)u_{3,j}(t)$$

by choosing the basis $\{N_j\}_{j=1,\ldots,m}$ to be finite element shape functions built from cubic order Hermite splines (Figure 8.5).

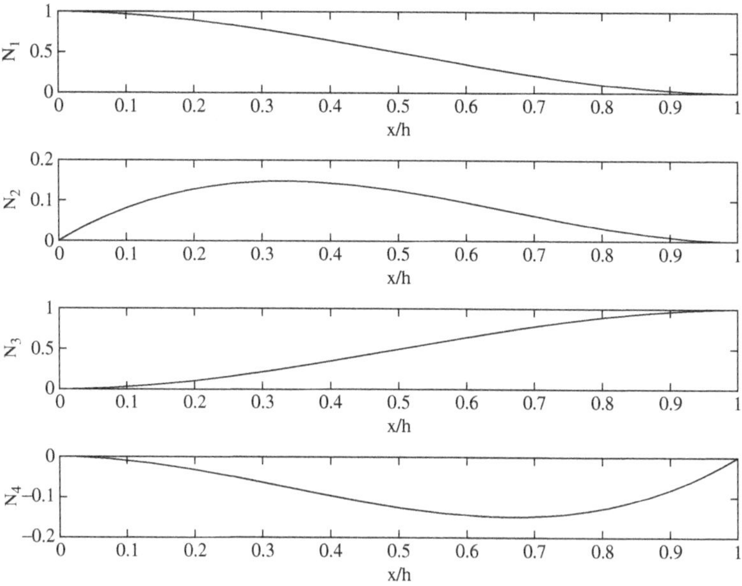

Figure 8.5 Conventional beam finite element functions over element e_1 with $h_1 = h$

In this example we consider a cantilevered beam which is fixed at $x_1 = 0$ and free at $x_1 = L$. Modifications of the discussion below for other boundary conditions are straightforward. As in Example 8.3.8, we partition the domain $[0, L]$ into elements $e_\ell := [x_{1,\ell-1}, x_{1,\ell}]$ for $\ell = 1, \ldots, n_d$ where $\{x_{1,\ell}\}_{\ell=0,\ldots,n_d}$ are the nodes or breakpoints of the spline basis. The constant h_ℓ denotes the width of element e_ℓ. We define the local cubic polynomials $N_{1,\ell}, N_{2,\ell}, N_{3,\ell}, N_{4,\ell}$ over each element e_ℓ for $\ell = 1, \ldots, n_d$. Figure 8.6 illustrates a collection of nodes, finite elements, and finite element shape functions defined over a typical beam.

(Continued)

Example 8.3.9 (Continued)

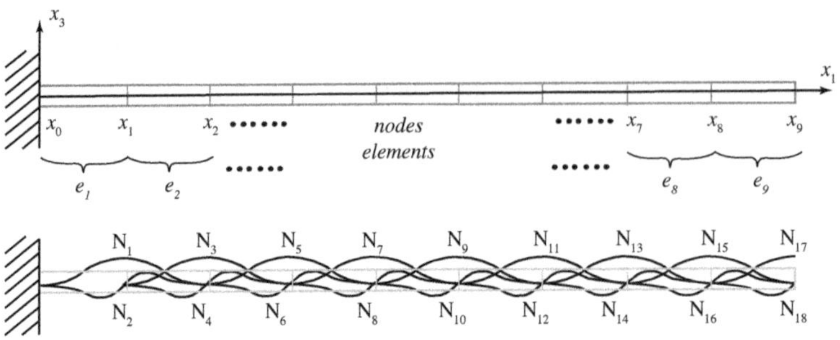

Figure 8.6 Nodes, elements, and beam finite element basis defined over a beam

The cubic Hermite polynomials that define the finite element shape functions satisfy the following four interpolatory conditions over element e_ℓ,

$$
\begin{aligned}
&(1)\quad N_{1,\ell}(x_{1,\ell-1}) = 1 \quad \frac{dN_{1,\ell}}{dx_1}(x_{1,\ell-1}) = 0 \quad N_{1,\ell}(x_{1,\ell}) = 0 \quad \frac{dN_{1,\ell}}{dx_1}(x_{1,\ell}) = 0,\\
&(2)\quad N_{2,\ell}(x_{1,\ell-1}) = 0 \quad \frac{dN_{2,\ell}}{dx_1}(x_{1,\ell-1}) = 1 \quad N_{2,\ell}(x_{1,\ell}) = 0 \quad \frac{dN_{2,\ell}}{dx_1}(x_{1,\ell}) = 0,\\
&(3)\quad N_{3,\ell}(x_{1,\ell-1}) = 0 \quad \frac{dN_{3,\ell}}{dx_1}(x_{1,\ell-1}) = 0 \quad N_{3,\ell}(x_{1,\ell}) = 1 \quad \frac{dN_{3,\ell}}{dx_1}(x_{1,\ell}) = 0,\\
&(4)\quad N_{4,\ell}(x_{1,\ell-1}) = 0 \quad \frac{dN_{4,\ell}}{dx_1}(x_{1,\ell-1}) = 0 \quad N_{4,\ell}(x_{1,\ell}) = 0 \quad \frac{dN_{4,\ell}}{dx_1}(x_{1,\ell}) = 1,
\end{aligned}
$$

for $\ell = 1, \ldots, n_d$. These four interpolatory conditions can be used to solve for the form of the cubic polynomials that define the beam shape functions over each element e_ℓ. For simplicity, suppose that the grid is uniform so that $h_\ell := h$ for all $\ell = 1, \ldots, n_d$. The solution for the shape functions $N_{1,1}, N_{2,1}, N_{3,1}$, and $N_{4,1}$ defined over the first element e_1 is given by [11]

$$
\begin{aligned}
N_{1,1}(x_1) &= 1 - 3\left(\frac{x_1}{h}\right)^2 + 2\left(\frac{x_1}{h}\right)^3,\\
N_{2,1}(x_1) &= x_1 - 2h\left(\frac{x_1}{h}\right)^2 + h\left(\frac{x_1}{h}\right)^3,\\
N_{3,1}(x_1) &= 3\left(\frac{x_1}{h}\right)^2 - 2\left(\frac{x_1}{h}\right)^3,\\
N_{4,1}(x_1) &= -h\left(\frac{x_1}{h}\right)^2 + h\left(\frac{x_1}{h}\right)^3.
\end{aligned}
$$

All the other shape functions $\{N_{1,\ell}, N_{2,\ell}, N_{3,\ell}, N_{4,\ell}\}_{\ell=2,\ldots,n_d}$ can be obtained by translation of these four functions. By choosing $N = N_i$ and substituting the Galerkin approximation $u_3 = N_j u_{3,j}$ in Equation 8.16, we obtain a set of governing equations

$$m_{ij}\frac{d^2u_j}{dt^2}(t) + k_{ij}u_j(t) - SN_i|_0^L + \mathcal{M}\frac{\partial N}{\partial x_1}|_0^L = b_i V(t)$$

where $m_{ij} = \int_0^L \rho_m A N_i N_j dx_1$, $k_{ij} = \int_0^L C_{11}^E I \frac{\partial^2 N_i}{\partial x_1^2}\frac{\partial^2 N_j}{\partial x_1^2} dx_1$, and $b_i = \int_0^L \frac{Ke_{31}\chi_{[a,b]}}{t_p}\frac{\partial^2 N_i}{\partial x_1^2} dx_1$. The global mass, stiffness, and control influence matrices m_{ij}, k_{ij}, b_i are again usually built by assembly of the elemental mass, stiffness, and control influence matrices $m_{ij}^{e_\ell}, k_{ij}^{e_\ell}, b_i^{e_\ell}$ that are obtained by restricting integration to a single finite element e_ℓ. For the mass matrix, we have

$$m_{ij} := \int_0^L \rho_m A N_i N_j dx_1 = \sum_{\ell=1}^{n_d}\int_{x_{l-1}}^{x_l} \rho_m A N_i N_j dx_1 = \sum_{\ell=1}^{n_d} m_{ij}^{e_l}.$$

The global stiffness matrix k_{ij} is computed similarly in terms of the elemental stiffness matrices. The elemental mass, stiffness, and control influence matrices for Bernoulli–Euler beam are given as follows:

$$m_{ij}^{e_\ell} = \begin{bmatrix} \int_{e_\ell}\rho_m A N_{1,\ell}N_{1,\ell}dx_1 & \cdots & \int_{e_\ell}\rho_m A N_{1,\ell}N_{4,\ell}dx_1 \\ \vdots & \cdots & \vdots \\ \int_{e_\ell}\rho_m A N_{4,\ell}N_{1,\ell}dx_1 & \cdots & \int_{e_\ell}\rho_m A N_{4,\ell}N_{4,\ell}dx_1 \end{bmatrix}$$

$$= \frac{\rho_m A h_n}{420}\begin{bmatrix} 156 & 22h & 54 & -13h \\ 22h & 4h^2 & 13h & -3h^2 \\ 54 & 13h & 156 & -22h \\ -13h & -3h^2 & -22h & 4h^2 \end{bmatrix},$$

$$k_{ij}^{e_\ell} = \begin{bmatrix} \int_{e_n} C_{11}^E I N''_{1,\ell}N''_{1,\ell}dx_1 & \cdots & \int_{e_n} C_{11}^E I N''_{1,\ell}N''_{4,\ell}dx_1 \\ \vdots & \cdots & \vdots \\ \int_{e_n} C_{11}^E I N''_{4,\ell}N''_{1,\ell}dx_1 & \cdots & \int_{e_\ell} C_{11}^E I N''_{4,\ell}N''_{4,\ell}dx_1 \end{bmatrix}$$

$$= \frac{C_{11}^E I}{h^3}\begin{bmatrix} 12 & 6h & -12 & 6h \\ 6h & 4h^2 & -6h & 2h^2 \\ -12 & -6h & 12 & -6h \\ 6h & 2h^2 & -6h & 4h^2 \end{bmatrix},$$

$$b_i^{e_\ell} = \begin{bmatrix} \int_{e_\ell}\frac{Ke_{31}\chi_{[a,b]}}{t_p}N''_{1,\ell}dx_1 \\ \vdots \\ \int_{e_\ell}\frac{Ke_{31}\chi_{[a,b]}}{t_p}N''_{4,\ell}dx_1 \end{bmatrix} = \frac{Ke_{31}}{t_p}\begin{bmatrix} 0 \\ 1 \\ 0 \\ -1 \end{bmatrix}.$$

(Continued)

Example 8.3.9 (Continued)

The elemental matrices k_{ij}^{el}, m_{ij}^{el} above are documented in classical texts on structural dynamics [11] or finite element texts [19]. As before, the global mass, stiffness, and control matrices are constructed by the assembly process. Detailed examples of the matrix assembly process can be found in [11, 19], or [51].

Upon completion of the assembly process, we obtain the equations

$$m_{ij}\ddot{u}_{3,j}(t) + c_{ij}\dot{u}_{3,j}(t) + k_{ij}u_{3,j}(t) = b_i V(t),$$

for $i = 1, \dots, n_d$ that govern the degrees of freedom $u_{3,j}(t)$ for $j = 1, \dots n_d$ in the Galerkin approximation $u_3(t, x_1) = N_j(x_1)u_{3,j}(t)$.

Table 8.2 A piezoelectric composite beam material and system properties.

Parameter	Description	Value	Units
$C_{11}^{E,s}$	Substrate Young's modulus	$6.9E10$	N/m^2
$C_{11}^{E,p}$	piezoelectric Young's modulus	$6.3E10$	N/m^2
ρ_m^s	Substrate density	2700	kg/m^2
ρ_m^p	Piezoelectric density	7800	kg/m^2
W	Width	0.01	m
L	Length	0.1	m
a	patch coordinate, left	0.4*L	m
b	patch coordinate, right	0.6*L	m
t_s	Substrate thickness	0.1	m
t_p	Piezoelectric thickness	$1.0E-7$	m
$d_{33} = e_{33}C_{33}^E$	Piezoelectric constant	$400E-12$	C/N
m_t	Tip mass	0	kg
L	Length	0.1	m
ξ_k	Modal damping factor, $k = 1 \dots n$	0.01	Unitless

In Figure 8.7, the numerically estimated natural frequencies $\omega_{n,k}^{n_d}$ and the analytic natural frequencies $\omega_{n,k}^a$ are compared for modes $k = 1, \dots, 10$ and model dimension $n_d = 10, 20, 40, 60, 80, 100$. Qualitatively, the results resemble those of Example 8.3.8. First, note that the thickness of the piezoelectric patch has been selected to be negligibly thin, $t_p = 1.E-7$ m, and that the tip

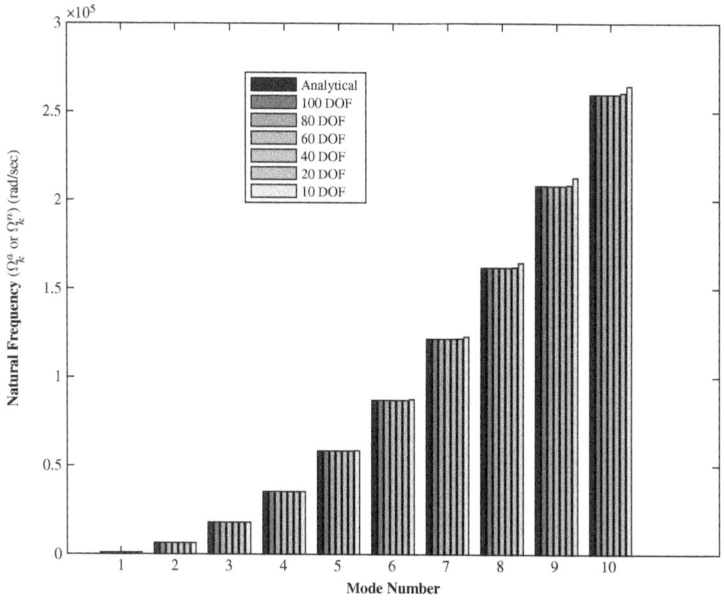

Figure 8.7 Comparison of numerically computed frequencies $\omega_{n,k}^{n_d}$ and analytic frequencies $\omega_{n,k}^{a}$ for modes $k = 1, \dots, 10$ and degrees of freedom $n_d = 10, 20, 40, 60, 80, 100$.

mass $m_t = 0$, in this example. The piezoelectric patch does not significantly change the mass and stiffness matrices, and we expect that the numerically calculated frequencies $\omega_{n,k}^{n_d}$ should closely match the analytic values $\omega_{n,k}^{a}$ as the dimension of the model increases. This is indeed the case in the figure. A careful inspection of the figure also illustrates that for a given mode k, the convergence of the numeric estimates $\omega_{n,k}^{n_d}$ approach the analytic value $\omega_{n,k}^{a}$ from above as n_d increases. Also, the error in approximating $\omega_{n,k}^{a}$ by $\omega_{n,k}^{n_d}$ monotonically increases when n_d is fixed and k increases. These are the same trends as noted in Example 8.3.8.

The results of transient numerical simulation of the equations obtained from the Galerkin approximation are shown in Figure 8.8. The tip displacement $u_3(t, L)$ in (b) is the trace of the surface plot in (a) at $x_1 = L$. As in our study of the axial piezoelectric specimen, the response quickly converges to a steady state oscillation. The large displacement of the free end of the beam can be understood by recalling that the piezoelectric patch applies an external load to the host material that is equivalent to concentrated moments at

(Continued)

Example 8.3.9 (Continued)

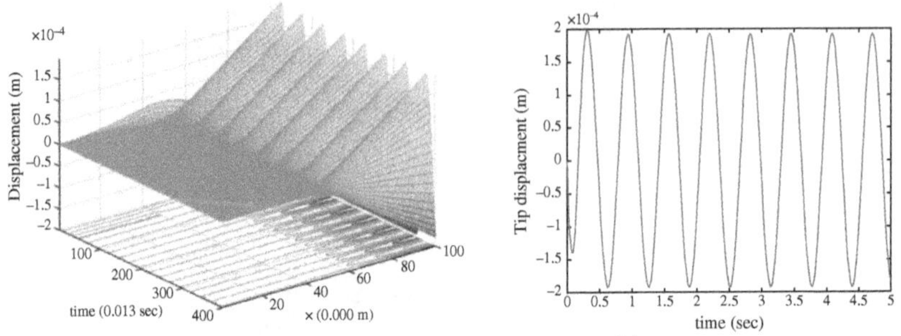

(a) Displacement $u_3\ (t, x_1)$ (b) Tip displacement $u3\ (t, L)$

Figure 8.8 Transient response of composite piezoelectric beam specimen

the edges of patch. The portion of the beam that is to the right of the patch essentially rotates as a rigid body due to the change in slope $\frac{\partial u_3}{\partial x_1}(t, b)$ at the right side of the piezoelectric patch.

As in Example 8.3.8 The frequency response function from a harmonic voltage input $V(t) := V_0 e^{\hat{j}\varpi t}$ to the harmonic output displacement $u_3(t, L)$ of the tip at $x_3 = L$ of the beam is given by

$$u_3(t, L) = N_i(L)\psi_{i,\ell} H_{\ell,j}(\hat{j}\varpi) b_j V_0 e^{\hat{j}\varpi t},$$
$$= N_i(L)\mathbb{H}_{i,j}(\hat{j}\varpi) b_j V_0 e^{\hat{j}\varpi t} = H(\hat{j}\varpi)) V_0 e^{\hat{j}\varpi t},$$

with the summation over $i, j, \ell = 1, \ldots, n_d$, and

$$\mathbb{H}_{i,j}(\hat{j}\varpi) := \frac{\psi_{i,\ell}\psi_{j,\ell}}{(1 - r_\ell^2) + 2\xi_\ell r_\ell \hat{j}} \frac{1}{k_\ell},$$

with the summation convention only over $\ell = 1, \ldots, n_d$. The Bode plot of $H(\hat{j}\varpi) := N_i(L)\mathbb{H}_{i,j}(\hat{j}\varpi) b_j$ consists of the magnitude plot of 20 log $|H(\hat{j}\varpi)|$ in decibels (db) and the plot of the phase $\arg(H(\hat{j}\varpi))$ (Figure 8.9).

Again, we see excellent agreement between the two Bode plots of the models having dimension $n_d = 10$ and $n_d = 100$. The magnitude plots are in particularly good agreement in the lower range of the driving frequency Ω. Discrepancies in the plots of the magnitude of the frequency response functions becomes noticeable starting when the driving frequency Ω enters the range from $\varpi = 10^4$ to $\varpi = 10^5$ rad/sec.

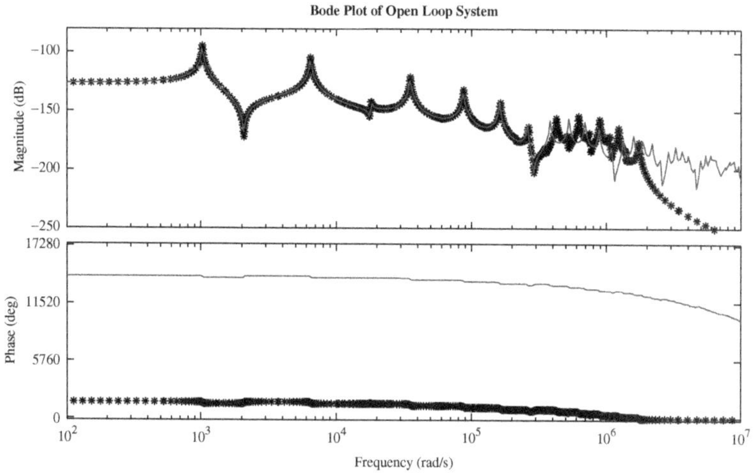

Figure 8.9 Comparison of Bode plot for $n_d = 10$ and $n_d = 100$ degrees of freedom in the Galerkin approximation $u_3(t, x_1) = N_j(x_1)u_{3,j}(t)$ for $j = 1, \ldots, n_d$.

In this example we study the construction of approximations for a composite piezoelectric beam with an attached shunt capacitor circuit.

Example 8.3.10 The equations governing a Galerkin approximation of the base excited, piezoelectric composite beam with a shunt capacitor can be derived in a manner similar to the approach taken in Example 8.3.2. In this case, the kinetic energy is expressed as

$$\mathcal{T} = \frac{1}{2}\int_0^L \rho_m A(\dot{u}_3 + \dot{z})^2 dx_1.$$

where $z(t)$ is the prescribed base motion of the piezoelectric composite. We introduce the Galerkin approximation $u_3(t, x_1) = N_j(x_1)u_{3,j}(t)$. Since the variation of the displacement $\delta u_3(t, x_1) = N_i(x_1)\delta u_{3,i}(t)$, the variation of the kinetic energy then becomes

$$\delta \int_{t_0}^{t_1} \mathcal{T} dt = \int_{t_0}^{t_1} \int_0^L \rho_m A(\dot{u}_3 + \dot{z})\delta\dot{u}_3 dx_1 dt,$$

(Continued)

Example 8.3.10 (Continued)

$$= \int_0^L \left\{ \rho_m A(\dot{u}_3 + \dot{z})\delta u_3\big|_{t_0}^{t_1} - \int_{t_0}^{t_1} \rho_m A(\ddot{u}_3 + \ddot{z})\delta u_3 dt \right\} dx_1,$$

$$= -\int_{t_0}^{t_1} \left\{ m_{ij}\ddot{u}_{3,j} + \underbrace{\int_0^L \rho_m A N_i dx_1 \ddot{z}}_{P_i} \right\} \delta u_{3,i} dt,$$

$$= -\int_{t_0}^{t_1} \{m_{ij}\ddot{u}_{3,j} + P_i\ddot{z}\}\delta u_{3,i} dt.$$

We assume that the system is subject to viscous damping that is represented in terms of a damping matrix c_{ij}. The equations governing the Galerkin approximation can be written in the form

$$m_{ij}\ddot{u}_j(t) + c_{ij}\dot{u}_j(t) + k_{ij}u_j(t) = b_i V(t) - P_i\ddot{z}(t).$$

The constants m_{ij}, k_{ij} and b_i have been defined in Equations 8.17, 8.18, and 8.19. In contrast to the previous examples, however, the voltage $V(t)$ is not prescribed in this example. Rather, it is an unknown, and its variation δV is not equal to zero in general. We will require an additional equation to obtain a well-posed problems in terms of the unknown displacement field $u_3(t, x_1)$ and unknown voltage $V(t)$. The current $i(t)$ out of the electrode on the bottom surface of the piezoelectric is given by

$$i(t) := -\dot{q}(t) = -\int_{A_e} \dot{\overline{\sigma}} da = \int_{A_e} \dot{D}_3 n_3 da = -\int \dot{D}_3 da$$

$$= \int_{A_e} e_{31} x_3 N_j'' dx_2 dx_1 \dot{u}_{3,j} - \int_{A_e} \frac{\epsilon_{33}\chi_{[a,b]}}{t_p} dx_2 dx_1 \dot{V}$$

$$= -\left\{ W\left(\frac{t_s}{2} + t_p\right) \int_0^L e_{31}\chi_{[a,b]} N_j''(x_1) dx_1 \dot{u}_j \right.$$

$$\left. + \int_a^b \int_{-W/2}^{W/2} \frac{\epsilon_{33}\chi_{[a,b]}}{t_p} dx_2 dx_1 \dot{V} \right\}.$$

We can check consistency by calculating the current $\tilde{i}$ out of the electrode on the top surface of the piezoelectric. Following the same essential steps, we have

$$\tilde{i}(t) = W\frac{t_s}{2} \int_0^L e_{31}\chi_{[a,b]} N_j''(x_1) dx_1 \dot{u}_j + \int_a^b \int_{-W/2}^{W/2} \frac{\epsilon_{33}\chi_{[a,b]}}{t_p} dx_2 dx_1 \dot{V}$$

Carefully note that we do not have $i(t) = -\tilde{i}(t)$, as expected. This difference can be attributed to the fact that the constant electric field assumption $E_3 = \mp V/t_p$ does not exactly satisfy the initial-boundary value problem of linear piezoelectricity. However, we choose the approximate current

$$\begin{aligned} i_a &:= \frac{i(t) - \tilde{i}(t)}{2} \\ &= -\left\{ W(t_s + t_p) \int_0^L e_{31}\chi_{[a,b]} N_j''(x_1) dx_1 \dot{u}_j \right. \\ &\quad \left. + \int_a^b \int_{-W/2}^{W/2} \frac{\epsilon_{33}\chi_{[a,b]}}{t_p} dx_2 dx_1 \dot{V} \right\}. \end{aligned}$$

Recall from Section 6.3 that $\kappa = \kappa_T - \kappa_B$ is the area moment of the cross section of the piezoelectric composite beam. In this example we have $\kappa_T = -\kappa_B = \frac{1}{2}Wt_p(t_s + t_p)$. A final equation satisfied by the approximate current i_a is then

$$0 = b_j \dot{u}_{3,j} + C_p \dot{V} + i_a \tag{8.35}$$

with $C_p := \frac{\epsilon_{33}W(b-a)}{t_p}$ and $b_j := \int_0^L \frac{\kappa e_{31}\chi_{[a,b]}}{t_p} N_j'' dx_1$. The shunt circuit containing the capacitor gives rise to the well known identity $i_a = C_s\dot{V}$, and the system of equations satisfied by the base excited, piezoelectric composite beam with a shunt capacitor are

$$\begin{aligned} m_{ij}\ddot{u}_{3,j}(t) + c_{ij}\dot{u}_{3,j}(t) + k_{ij}u_{3,j}(t) &= b_i V(t) - P_i \ddot{z}(t) \\ (C_p + C_s)\dot{V}(t) + b_j \dot{u}_{3,j}(t) &= 0. \end{aligned}$$

In the last example the approximate equations of motion for the capacitively shunted piezoelectric beam are derived by using an expression for the sensor Equation 8.35 from the piezoelectric constitutive laws. In the next example, we derive the same set of equations using Hamilton's Principle for electromechanical systems.

Example 8.3.11 The electromechanical potential $\mathcal{V}_{em}$ for the system considered in Example 7.4.2 is given by

$$\mathcal{V}_{em} = \mathcal{V}_H - \mathcal{V}_e,$$

(Continued)

Example 8.3.11 (Continued)

with $\mathcal{V}_{\mathcal{H}}$ defined in Example 7.4.2, $\mathcal{V}_m = 0$, and $\mathcal{V}_e = \frac{1}{2}C_s V^2$. As shown in Example 7.4.2, the stationarity of $\mathcal{V}_{em}$ leads to an variational expression that is linear in the variations δu_3 and δV:

$$
\begin{aligned}
0 = \int_{t_0}^{t_1} \int_0^L \Bigg\{ -\rho_m A\left(\frac{\partial^2 u_3}{\partial t^2} + \ddot{z}\right) - \frac{\partial^2}{\partial x_1^2}\left(C_{11}^E I \frac{\partial^2 u_3}{\partial x_1^2}\right) \\
+ \frac{\partial^2}{\partial x_1^2}\left(\frac{K e_{31}\chi_{[a,b]}}{t_p}\right) V \Bigg\} \delta u_3 dx_1 dt \\
+ \int_{t_0}^{t_1} \left\{ \int_0^L \frac{K e_{31}\chi_{[a,b]}}{t_p}\frac{\partial^2 u_3}{\partial x_1^2} dx_1 + (C_p + C_s)V \right\} \delta V\, dt \\
+ \int_{t_0}^{t_1} \left\{ -C_{11}^E I \frac{\partial^2 u_3}{\partial x_1^2} + \frac{K e_{31}\chi_{[a,b]}V}{t_p} \right\} \delta\left(\frac{\partial u_3}{\partial x_1}\right)\Bigg|_0^L dt \\
+ \int_{t_0}^{t_1} \left\{ \frac{\partial}{\partial x_1}\left(C_{11}^E I \frac{\partial^2 u_3}{\partial x_1^2}\right) - \frac{\partial}{\partial x_1}\left(\frac{K e_{31}\chi_{[a,b]}V}{t_p}\right) \right\} \delta u_3\big|_0^L dt
\end{aligned}
$$

When the Galerkin approximation is given by $u_3(t, x_1) = N_j(x_1)u_{3,j}(t)$ with the summation over $j = 1, \dots, n_d$, the variation $\delta u_3(t, x_1) := N_j(x_1)\delta u_{3,j}(t)$ since each $N_j(x_1)$ is a prescribed function. We substitute the Galerkin approximation $u_3 = N_j u_{3,j}$ and variations $\delta u_3 = N_i \delta u_{3,i}$ into the variational equations above, integrate by parts, and then enforce the boundary conditions. Then the Galerkin approximate solution $u_3 = N_j u_{3,j}$ must satisfy the variational statement

$$
\begin{aligned}
0 = \int_{t_0}^{t_1} \{m_{ij}\ddot{u}_{3,j}(t) + k_{ij}u_{3,j}(t) - b_i V(t) + P_i \ddot{z}(t)\}\delta u_{3,i}(t) dt \\
+ \int_{t_0}^{t_1} \{b_k u_{3,k}(t) + (C_p + C_s)V\}\delta V(t) dt
\end{aligned}
$$

for all variations $\delta u_{3,i}$ and δV. These equations are identical to those derived in Example 8.3.10 when the damping matrix $c_{ij} = 0$. If the virtual work of the nonconservative forces is updated in this example to include viscous damping, we obtain the exact same equations as in Example 8.3.10.

In this last example we return the the problem studied in Example 6.4.2 and discuss the results of a Galerkin approximation of the equations of motion.

Example 8.3.12 The equations that govern the displacement $u_3(t,x_1)$ of the composite piezoelectrically actuated beam are derived in Example 6.4.2. In Example 6.4.2 the domain $\Omega = [0,L]$ is subdivided into five subdomains $\Omega = \cup_{i=1}^{5}\Omega_i$ whose boundaries lie at the locations where the material properties are discontinuous. We have

$$\Omega_1,\ldots,\Omega_5 = [0,a_1],[a_1,b_1],[b_1,a_2],[a_2,b_2],[b_2,L].$$

As shown in Example 6.4.2, it is then possible to define five local equations of motion over each Ω_i, each having constant material properties, and the entire collection is coupled by 20 boundary conditions that are summarized in the table in Example 6.4.2. Reference [27] introduces an approximation method for the solution of these five, coupled evolutionary partial differential equations and their associated boundary conditions. In the formulation each solution $u_{3,i}$ for $i = 1,\ldots,5$ is assumed to have the form

$$u_{3,i} := T_i(t)\psi_i(x_1).$$

The eigenproblem for the partial differential equation that holds over the ith subdomain Ω_i of the beam is derived by substituting separable form of $u_{3,i}$ into the ith equations of motion when the control influence term is set to zero. The result is that

$$-\rho_{m,i}A_i\frac{\frac{d^2T}{dt^2}}{T} = C^E_{11,i}I_i\frac{\frac{d^4\psi_i}{dx_1^4}}{\psi_i} = \hat{c}_i\rho_{m,i}A_i.$$

The spatial eigenproblem then results when we consider the rightmost equation above written in the form

$$\frac{d^4\psi_i}{dx_1^4} - \omega_i^4\psi_i = 0,$$

with $\omega_i = \frac{\rho_{m,i}A_i}{C^E_{11,i}I_i}\hat{c}_i$. Finally, the general form of the eigenfunction ψ_i takes the familiar modal expression

$$\psi_i(x_1) = a_{i,1}\sin\omega_i x_1 + a_{i,2}\cos\omega_i x_1 + a_{i,3}\sinh\omega_i x_1 + a_{i,4}\cosh\omega_i x_1.$$

It must be noted that the above equation is not in terms of the summation convention. The function ψ_i is the first mode of the eigenproblem over the domain Ω_i, and each summand above contains a single term. Reference [27] gives a detailed description of the rather lengthy process to impose the 20 boundary conditions on the solutions $u_i(t,x_1)$ over $\omega_{n,i}$ for $i = 1,\ldots,5$.

(Continued)

Example 8.3.12 (Continued)

Table 8.3 Material and system properties for the piezoelectric composite beam with two actuating patches in 6.4.2.

Length of the beam	**0.305 m**
Material of the beam	Brass
Young's modulus of the substrate	100 GPa
Density of the substrate	8430 kg=m3
Thickness of the substrate	0.832 mm
Width of the substrate	19.5 mm
Type of piezo-ceramic	PZT-5A
Density of the piezo-ceramic	7800 kg=m3
Young's modulus of the piezo-ceramic	62 GPa
Thickness of the piezo-ceramic	0.191 mm
$0, a_1, b_1, a_2, b_2, L$	$0, 0.02, 0.04, 0.265, 0.285, 0.305$ m

The experiment was conducted in [27] using the configuration shown in Figure 8.10. The composite piezoelectric beam was subject to two sinusoidal input voltages where the phase difference between the inputs was fixed for a particular trial. Measurements of the positions and velocities along the beam were recorded using the Polytec PSV-400 ™scanning laser vibrometer. Figure 8.11 depicts a comparison of the numerical prediction of the traveling waves generated by clamped-clamped boundary conditions with experiment. We have traveling waves in (a) generated by the model and in (b) the experimentally observed traveling waves. Good correlation is evident between the two figures, although the actual experimental values in (b) are more highly damped than is predicted by numerical simulation in (a). Figure 8.12 illustrates the comparison of traveling waves generated by clamped-free boundary conditions. The predictions of the numerical approximation appear in (a), while the experimentally observed waves are shown in (b). Likewise, the locations of the peaks and there frequencies agree well in (a) and (b), but the amplitude of vibration in the experimental results are noticeably smaller. Thus, again, the damping is underestimated in the numerical approximation. Figure 8.13 shows that for free-free boundary conditions, the frequency,

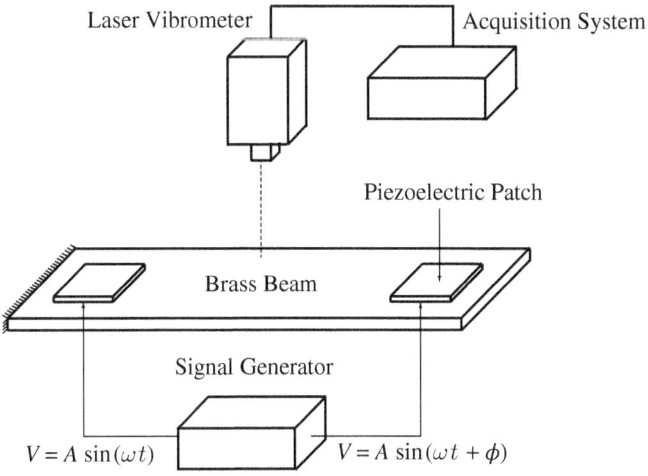

Figure 8.10 Experimental Setup, [27] Source: Vijaya Venkata Malladi / https://doi.org/10.1177/1045389X16679284

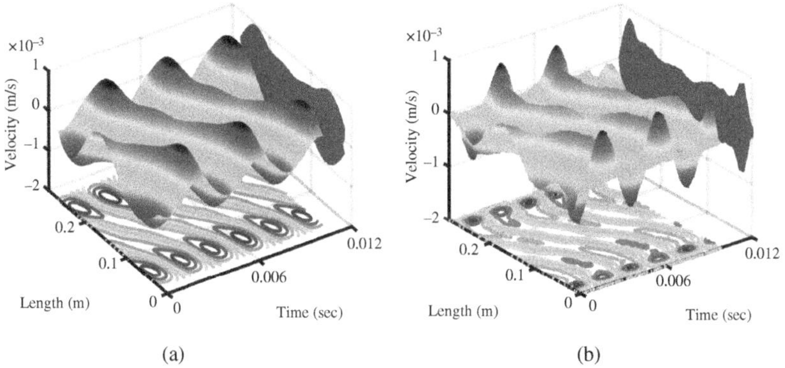

Figure 8.11 Clamped-clamped traveling waves, (a) analytical, (b) experimental [27].

phase, and amplitude are quite similar between the numerical simulation in (a) and the experimental observations plotted in (b). Figure 8.14 depicts how the traveling waves vary as the excitation frequency ranges from 0 Hz to 350 Hz for clamped-clamped boundary conditions in (a), and clamped-free boundary conditions in (b). Figure 8.15 shows the envelopes of the traveling waves produced by (a) clamped-clamped boundary conditions driven at 285 Hz and (b) clamped-free boundary conditions excited at 241 Hz.

(Continued)

Example 8.3.12 (Continued)

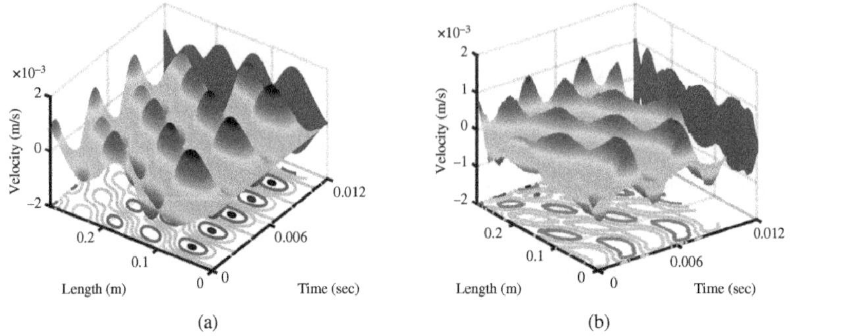

Figure 8.12 Clamped-free traveling waves, (a) analytical, (b) experimental [27].

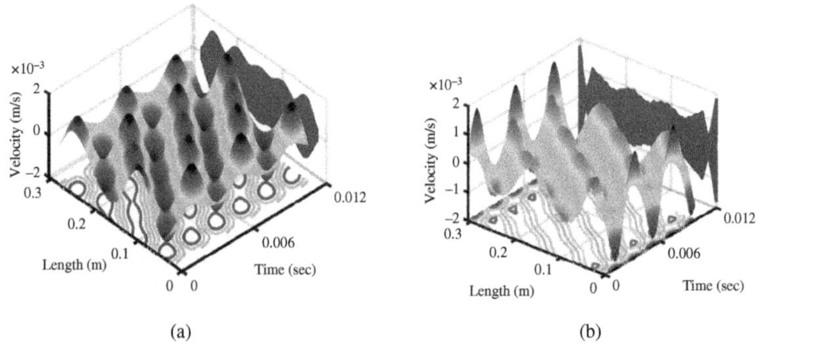

Figure 8.13 Free-free traveling waves, (a) analytical, (b) experimental [27].

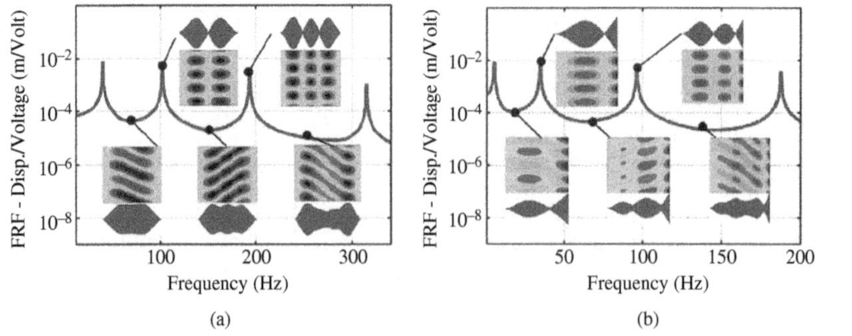

Figure 8.14 Analytical estimates of traveling waves, (a) clamped-clamped, (b) clamped-free [27].

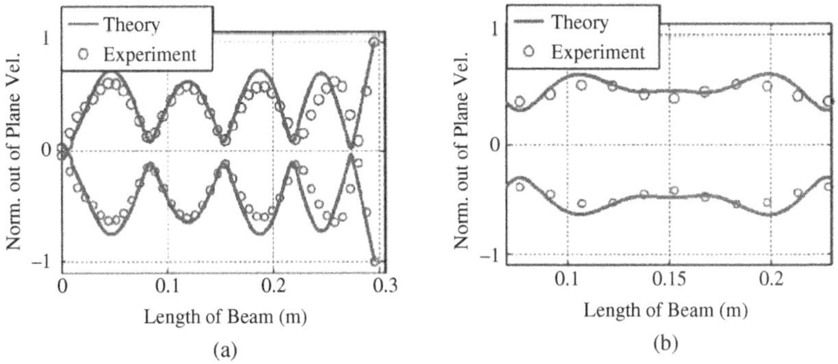

Figure 8.15 Envelopes of traveling waves, (a) clamped free, 285 Hz, (b) free-free, 241 Hz [27].

8.4 Problems

Problems 8.4.1 Construct a Galerkin approximation using Fourier modes for the resistively shunted composite piezoelectric beam studied in Section 7.4.

Problems 8.4.2 Construct a Galerkin approximation using finite element shape functions for the resistively shunted composite piezoelectric beam studied in Section 7.4.

Problems 8.4.3 Construct a Galerkin approximation using Fourier modes for the composite linearly piezoelectric rod that is driven by input base motion in Problem 6.5.2.

Problems 8.4.4 Construct a Galerkin approximation using finite element shape functions for the composite linearly piezoelectric rod that is driven by input base motion in Problem 6.5.2.

Problems 8.4.5 Construct a Galerkin approximation using Fourier modes for the composite linearly piezoelectric beam that is driven by input base motion in Problem 6.5.4.

Problems 8.4.6 Construct a Galerkin approximation using finite element shape functions for the composite linearly piezoelectric beam that is driven by input base motion in Problem 6.5.4.

Problems 8.4.7 Construct a Galerkin approximation using Fourier modes for the composite linearly piezoelectric rod with a tip mass in Problem 6.5.3.

Problems 8.4.8 Construct a Galerkin approximation using finite element shape functions for the composite linearly piezoelectric rod with a tip mass in Problem 6.5.3.

Problems 8.4.9 Construct a Galerkin approximation using Fourier modes for the composite linearly piezoelectric beam with a tip mass in Problem 6.5.3.

Problems 8.4.10 Construct a Galerkin approximation using finite element shape functions for the composite linearly piezoelectric beam with a tip mass in Problem 6.5.3.

Supplementary Material

S.1 A Review of Vibrations

Since this book is intended to be self-contained, several chapters introduce fundamental principles that are essential to a formulation of linear piezoelectricity. Chapters 2, 3, and 4 review the requisite background and supporting principles from mathematics, continuum mechanics, and continuum electrodynamics, respectively. In this section we review topics from the study of vibrations of linear structures that are used throughout Chapters 5, 6, 7, and 8. This review is necessarily brief, and a full account can be found in any of the excellent texts on this classical topic in mechanics. See for example [21, 29, 30], or [11].

Our presentation of vibrations of mechanical systems begins with a review of single degree of freedom (SDOF) systems in Section S.1.1. Methods for solving for the transient and steady state solution are reviewed, as well as frequency response techniques. Section S.1.2 discusses continuous, or distributed parameter systems (DPS). These systems are governed by equations of motion that define evolutions in a state space of continuous functions, or distributed parameters. Finally, Section S.1.3 summarizes approaches for treating multi-degree of freedom (MDOF) systems. Approximation of solutions to DPS are an important source of MDOF systems covered in this text, so the theory for SDOF systems and DPS precedes the discussion of MDOF systems.

S.1.1 SDOF Systems

Most treatises on the vibrations of mechanical systems begin with the study of scalar, second order, linear, time invariant equations of motion. By convention, the equation of motion is expressed in the familiar form

$$
\begin{aligned}
m\ddot{u}(t) + c\dot{u}(t) + ku(t) &= f(t), \qquad (8.36)\\
u(0) &= u_0,\\
\dot{u}(0) &= v_0,
\end{aligned}
$$

Vibrations of Linear Piezostructures, First Edition. Andrew J. Kurdila and Pablo A. Tarazaga.

This Work is a co-publication between John Wiley & Sons Ltd and ASME Press.

for each $t \in \mathbb{R}^+$, where m is the mass, c is the damping coefficient, k is the stiffness, and $f(t)$ is the applied force. This equation is studied in detail in introductory texts on vibrations such as [11] or [21], and in courses on linear ordinary differential equations. Numerous techniques have been developed to solve this equation, and popular strategies that have found widespread use in the engineering community include time domain methods, Laplace transform techniques, and Fourier transform approaches. See [35] or [30] for a discussion of Laplace transform methods. A discussion of Fourier methods can also be found in [11] or [30].

In this section we review some standard methods for solving the SDOF equation in the time domain. In vibrations textbooks Equation 8.36 is usually transformed into the canonical form

$$\ddot{u}(t) + 2\xi\omega_n\dot{u}(t) + \omega_n^2 u(t) = \frac{1}{m}f(t) =: p(t) \tag{8.37}$$

that introduces the damping parameter ξ and natural frequency ω_n from the identities $\omega_n^2 = k/m$ and $2\xi\omega_n = c/m$. The general form of the solution for either Equation 8.36 or 8.37 is expressed as the sum $u(t) = u^h(t) + u^p(t)$ of the homogeneous solution $u^h(t)$ and the particular solution $u^p(t)$. The homogeneous solution solves the governing equation when $f(t) \equiv p(t) \equiv 0$. Any solution that satisfies either Equations 8.36 or 8.37 when $f(t)$ or $p(t)$ is nonzero is a particular solution.

Homogeneous Solutions

The homogeneous solution $u^h(t)$ can be obtained by assuming a solution of the form $\mathbb{C}e^{\eta t}$, substituting it into Equation 8.37 when $f(t) \equiv p(t) \equiv 0$, and solving the resulting characteristic polynomial $\eta^2 + 2\xi\omega_n\eta + \omega_n^2 = 0$ for the roots $\eta_1, \eta_2 = \xi\omega_n \pm \sqrt{\xi^2 - 1}\omega_n$. The qualitative nature of the homogeneous solution depends on the sign of the discriminant $\sqrt{\xi^2 - 1}$. We say that a solution or system is overdamped, critically damped, or underdamped when $\xi > 1$, $\xi = 1$, or $\xi < 1$, respectively. In this brief review we focus on underdamped systems. The reader will find an exposition on overdamped or critically damped systems in [11] or [21]. For an underdamped system we have $\xi < 1$, and the homogeneous solution can be written in any of the following equivalent forms

$$\begin{aligned} u^h(t) &= e^{-\xi\omega_n t}(\mathbb{C}_1 e^{\hat{j}\omega_d t} + \mathbb{C}_2 e^{-\hat{j}\omega_d t}), \\ &= e^{-\xi\omega_n t}(A\cos\omega_d t + B\sin\omega_d t), \\ &= Ce^{-\xi\omega_n t}\cos(\omega_d t + \alpha), \\ &= Ce^{-\xi\omega_n t}\sin(\omega_d t + \beta), \end{aligned}$$

in which $\mathbb{C}_1, \mathbb{C}_2$ are complex constants, A, B, C are real constants, α, β are phase angles, and $\omega_d := \sqrt{1 - \xi^2}\omega_n$ is the damped natural frequency. See [21] for a discussion of the relationships among these representations.

Particular Solutions and FRFs

The determination of the particular solution of Equation 8.37 depends on the specific form of the forcing function $p(t)$. Analytic methods for determining a particular solution when the $p(t)$ is a polynomial, sine, cosine, product of a polynomial and sine, product of a polynomial and cosine, or a linear combinations of these expressions are discussed in introductory texts on ordinary differential equations or linear vibrations [21].

In the field of vibrations, when the forcing function $p(t)$ is proportional to either $\cos \varpi t$ or $\sin \varpi t$, for some driving frequency ϖ, we say that the system is driven by harmonic excitation. We consider only the case when the forcing function $p(t)$ is a linear combination of harmonic functions in this section. An understanding of this case can be used in conjunction with Fourier series methods to solve for the particular solution for quite general forcing functions $f(t)$ [11].

The complex response method is the most popular technique discussed in nearly all engineering textbooks for determining a particular solution when the system is driven by harmonics. Suppose the amplitude of the harmonic input is P_0 and the driving frequency is ϖ. The complex response method considers two real-valued problems simultaneously,

$$\ddot{u}(t) + 2\xi\omega_n\dot{u}(t) + \omega_n^2 u(t) = P_0 \cos \varpi t, \tag{8.38}$$

$$\ddot{v}(t) + 2\xi\omega_n\dot{v}(t) + \omega_n^2 v(t) = P_0 \sin \varpi t. \tag{8.39}$$

The real variable $u(t)$ is the response to the input $p(t) := P_0 \cos(\varpi t)$, and the real variable $v(t)$ is the solution to the input $p(t) := P_0 \sin(\varpi t)$. A complex equation is obtained by multiplying Equation 8.39 by $\hat{j} := \sqrt{-1}$, adding the resulting equation to Equation 8.38, and defining the complex response $z(t) := u(t) + \hat{j}v(t)$ as the solution of the complex response equation

$$\ddot{z}(t) + 2\xi\omega_n\dot{z}(t) + \omega_n^2 z(t) = P_0 e^{\hat{j}\varpi t}. \tag{8.40}$$

If we solve for the complex response $z(t)$ in Equation 8.40, we obtain the particular solution of the original Equation 8.39 by setting the solution to be given by $u(t) = \mathrm{re}(z(t))$ or $v(t) = \mathrm{im}(z(t))$, respectively.

The solution of the complex response equation assumes that $z(t)$ has the form $z(t) = \mathbb{Z}e^{\hat{j}\varpi t}$ for some complex constant $\mathbb{Z}$. On substituting this expression in the complex response equation, we find the complex response amplitude $\mathbb{Z}$ in terms of the complex frequency response function (FRF) $H(\hat{j}\varpi)$,

$$\mathbb{Z} = H(\hat{j}\varpi)P_0 := \frac{1}{(\omega_n^2 - \varpi^2) + 2\xi\varpi\omega_n\hat{j}} P_0.$$

Physically, the complex frequency response function $H(\hat{j}\varpi)$ can be understood as the ratio of the complex response $\mathbb{Z}$ to the amplitude P_0 of the harmonic input.

Any complex variable such as $\mathbb{Z}$ or $H(\hat{j}\varpi)$ can be written in polar form. We have $\mathbb{Z} = |\mathbb{Z}|e^{\hat{j}\beta} = |H(\hat{j}\varpi)|P_0 e^{\hat{j}\beta}$ for some phase angle β,

$$|\mathbb{Z}| = |H(\hat{j}\varpi)|P_0 = \frac{1}{\sqrt{(\omega_n^2 - \varpi^2)^2 + (2\xi\varpi\omega_n)^2}} P_0,$$

$$\beta = -\arctan\frac{2\xi\varpi\omega_n}{\omega_n^2 - \varpi^2}.$$

The response v to a sinusoidal input having driving frequency ϖ and amplitude P_0 is then given by

$$v(t) = \mathrm{im}(z(t)) = \mathrm{im}(|\mathbb{Z}|e^{\hat{j}\beta}e^{\hat{j}\varpi t}) = |\mathbb{Z}|\mathrm{im}(e^{\hat{j}(\varpi t+\beta)}) = |H(\hat{j}\varpi)|P_0 \sin(\varpi t + \beta),$$
$$= \frac{1}{\sqrt{(\omega_n^2 - \varpi^2)^2 + (2\xi\varpi\omega_n)^2}} P_0 \sin(\varpi t + \beta).$$

Following similar reasoning, the response $u(t)$ to the harmonic input $P_0 \cos \varpi t$ is given by $u(t) = |H(\hat{j}\varpi)|P_0 \cos(\varpi t + \beta)$. It should also be noted that the complex response coefficient $\mathbb{Z}$ is often written as

$$\mathbb{Z} = H(\hat{j}\varpi)P_0 = \frac{1}{((1 - r^2) + 2\xi r\hat{j})}\frac{1}{\omega_n^2}P_0,$$

in terms of the ratio r of the driving freqency to natural frequency, $r := \varpi/\omega_n$. Correspondingly, we have

$$|\mathbb{Z}| = \frac{1}{\sqrt{(1 - r^2)^2 + (2\xi r)^2}}\frac{1}{\omega_n^2}P_0,$$

$$\beta = -\arctan\frac{2\xi r}{1 - r^2}.$$

We have chosen to define the FRF $H(\hat{j}\varpi)$ in terms of the solution of the complex response equation that is formed by the superposition of two real equations. This analysis is self-contained, direct, and intuitive. It is also popular to define the FRF in terms of the system transfer function for the equation

$$\ddot{u}(t) + 2\xi\omega_n\dot{u}(t) + \omega_n^2 u(t) = p(t).$$

When the equation above is subject to zero initial conditions, we have

$$H(s) := \frac{u(s)}{p(s)} = \frac{1}{s^2 + 2\xi\omega_n s + \omega_n^2},$$

and subsequently define the FRF as

$$H(\hat{j}\varpi) := H(s)|_{s=\hat{j}\varpi} = \frac{1}{(\omega_n^2 - \varpi^2) + 2\xi\varpi\omega_n\hat{j}} = \frac{1}{(1 - r^2) + 2\xi r\hat{j}}\frac{1}{\omega_n^2}. \quad (8.41)$$

This result is identical to that obtained by solving the complex response equation. The interpretation of $|H(\hat{j}\varpi)|$ is the same in either definition. It is the ratio of the amplitude of the harmonic output to the amplitude of the harmonic input when the driving frequency is ϖ.

The popularity of the complex frequency response function $H(\hat{j}\varpi)$ in vibrations applications can be attributed in part to its succinct representation of the steady state response for the important class of harmonic inputs. Another factor contributing to its popularity is its amenability for describing experimental results. The complex frequency response function is a convenient way to represent the steady state response when conducting an experiment where the steady state output is measured for several input driving frequencies. Such an experiment is referred to as a sine sweep, or sine dwell, test.

Total Solutions

The total solution $u(t) := u^h(t) + u^p(t)$ for the SDOF system in Equation 8.37 can be written as

$$u(t) = e^{-\xi\omega_n t}(A\cos\omega_d t + B\sin\omega_d t) + \frac{1}{\sqrt{(1-r^2)^2 + (2\xi r)^2}}\frac{1}{\omega_n^2}P_0 e^{\hat{j}\varpi t} \qquad (8.42)$$

when the system is driven by a harmonic input. As before, the solution above is interpreted as a complex response since $e^{\hat{j}\varpi t}$ is a complex function. We choose the real part to obtain the solution driven by $p(t) = P_0\cos\varpi t$, and we select the imaginary part to obtain the solution driven by $p(t) := P_0\sin\varpi t$. The constants A and B are determined by applying the initial conditions $u(0) = u_0$ and $\dot{u}(0) = v_0$ to Equation 8.42.

Generalizations of SDOF Systems

For the SDOF system, the trajectory $t \mapsto u(t)$ evolves or takes values in the real numbers $\mathbb{R}$. We say that the state space for the system is $\mathbb{R}$. Two classes of systems are usually studied in engineering vibrations that are generalizations of the SDOF case. It is common in applications that a mechanical system cannot be modeled using a single, time-varying parameter $u(t) \in \mathbb{R}$ for each $t \in \mathbb{R}^+$. Instead, an $n-$vector $u := \{u_1, \ldots, u_n\} \in \mathbb{R}^n$ is required to represent the physics at hand. That is, the trajectory $t \mapsto u(t)$ describing the dynamics of the system evolves in the state space $\mathbb{R}^n$. Such a system is known as a multi-degree of freedom (MDOF) system and is studied in Section S.1.3 [11, 21, 29].

It also occurs that the state $u(t)$ of a system at a fixed time $t \in \mathbb{R}^+$ is not a scalar in $\mathbb{R}$, nor a vector in $\mathbb{R}^n$. The state $u(t)$ at each fixed time t can be a *function* in some abstract space H of functions. In mechanical vibrations the space H contains functions of the a spatial variable $x \in \Omega$ where Ω is the spatial domain. In other words, the system trajectories $t \mapsto u(t)$ take values in a space H, and at each $t \in \mathbb{R}^+$ we have $u(t) := u(t,x)$ of the variable $x \in \Omega$. Systems of this type are known as distributed parameter systems (DPS). A DPS is an example of an infinite dimensional dynamic system, one whose state evolves in an infinite dimensional space H. Such systems usually are approximated to generate practical predictions of response, and these approximations take the form of MDOF systems. Since the approximation of DPS yield MDOF models, we discuss DPS next.

S.1.2 Distributed Parameter Systems

A DPS system contains at least one state that belongs to an infinite dimensional space. Many types of equations of motion can generate a DPS including integral equations, delay equations, or functional differential equations. Perhaps the most common sources of DPS in classical texts on vibrations are partial differential equations (PDEs). The following two examples are illustrative of this case.

Example S.1.1 Consider a linearly elastic axial rod of length L having density ρ, cross sectional area A, and elastic modulus C. Let $u(t,x)$ be the axial deformation at time $t \in \mathbb{R}^+$ at a location $x \in \Omega := [0,L]$ along the length of the rod. Suppose that the deformation $u(t,x)$ is driven by a force $f(t,x)$ that is applied along the axial direction of the rod at time $t \in \mathbb{R}^+$ and at a location $x \in \Omega$. It can be shown [11, 21] that the displacement $u(t,x)$ satisfies the partial differential equation

$$\rho A \frac{\partial^2 u}{\partial t^2}(t,x) = \frac{\partial}{\partial x}\left(AC \frac{\partial u}{\partial x}(t,x) \right) + f(t,x), \tag{8.43}$$

for all $t \in \mathbb{R}^+$ and $x \in \Omega$, subject to appropriate boundary conditions such as

$$\begin{aligned} u(t,0) &= 0, \\ u(t,L) &= 0, \end{aligned} \tag{8.44}$$

for all $t \in \mathbb{R}^+$, and to the initial conditions

$$\begin{aligned} u(0,x) &= \overline{u}_0(x), \\ \frac{\partial u}{\partial t}(0,x) &= \overline{v}_0(x), \end{aligned} \tag{8.45}$$

for all $x \in \Omega$. The governing PDE 8.43, boundary conditions 8.44, and initial conditions 8.45 define a DPS.

Example S.1.2 Consider a Bernoulli–Euler beam of length L having mass density ρ, cross sectional area A, and bending stiffness CI. Let $u(t,x)$ be the lateral displacement of the beam at time $t \in \mathbb{R}^+$ and location $x \in \Omega := [0,L]$ along its length. The functions $f(t,x)$, $\mathcal{S}(t,x)$, and $\mathcal{M}(t,x)$ denote the transverse force, shear force, and bending moment, respectively, at time

$t \in \mathbb{R}^+$ at a location $x \in \Omega$ along the length of the beam. It can be shown that the displacement $u(t,x)$ satisfies the partial differential equation

$$\rho A \frac{\partial^2 u}{\partial t^2}(t,x) = -\frac{\partial^2}{\partial x^2}\left(CI \frac{\partial^2 u}{\partial x^2}(t,x) \right) + f(t,x), \tag{8.46}$$

for all $t \in \mathbb{R}^+$ and $x \in \Omega$, subject to appropriate boundary conditions such as

$$\begin{aligned} u(t,0) &= 0, \quad \mathcal{S}(t,L) = 0, \\ \tfrac{\partial u}{\partial t}(t,0) &= 0, \quad \mathcal{M}(t,L) = 0, \end{aligned} \tag{8.47}$$

for all $t \in \mathbb{R}^+$, and to the initial conditions

$$\begin{aligned} u(0,x) &= \bar{u}_0(x), \\ \frac{\partial u}{\partial t}(0,x) &= \overline{v}_0(x), \end{aligned} \tag{8.48}$$

for all $x \in \Omega$. The governing PDEs in Equation 8.46, the boundary conditions in 8.47, and the initial conditions in 8.48 define a DPS.

The reader will recognize Example S.1.1 as the linearly elastic model corresponding the axial piezoelectric rod studied in Equation 8.3.5. Example S.1.2 is the linearly elastic beam model corresponding to the piezoelectric composite beam studied in Equation 8.3.6. Other choices of boundary conditions are discussed in [21] and [11]. In both Examples S.1.1 and S.1.2, the unknown displacement $u(t,x)$ depends on two independent variables $t \in \mathbb{R}^+$ and $x \in \Omega := [0,L]$. The expression $u(t,x)$ denotes the value of the function u at a specific $(t,x) \in \mathbb{R}^+ \times \Omega$. By convention, we denote by $u(t) := u(t,\cdot)$ the function over Ω at a particular time $t \in \mathbb{R}^+$. With this notation, the trajectory

$$t \mapsto u(t) := u(t,\cdot)$$

takes values in a collection H of functions defined over Ω.

In general then, we can represent either Example S.1.1 or S.1.2 in a common form where we seek $u(t) := u(t,\cdot) \in H$ that satisfies

$$\mathfrak{M}\frac{\partial^2 u}{\partial t^2}(t) + \mathfrak{K}u(t) = f(t) \tag{8.49}$$

for $t \in \mathbb{R}^+$, subject to suitable boundary and initial conditions. The space H is a family of functions defined over the spatial domain Ω. We also interpret the inhomogeneous forcing function $f(t) := f(t,\cdot) \in H$. In Examples S.1.1 and S.1.2, the domain Ω is just $\Omega = [0,L] \subset \mathbb{R}$. The symbols $\mathfrak{M}$ and $\mathfrak{K}$ represent differential operators of the spatial variable $x \in \Omega$. In Example S.1.1, the operator $\mathfrak{M}$ is a multiplication operator. It is an order zero differential operator that depends on the spatial variable $x \in \Omega$, and $\mathfrak{K}$ is a second order differential operator in the spatial variable $x \in \Omega$,

$$\mathfrak{M}(\cdot) := \rho(x)A(x)(\cdot),$$

$$\mathfrak{K}(\cdot) := -\frac{\partial}{\partial x}\left(A(x)C(x)\frac{\partial}{\partial x}(\cdot)\right).$$

In Example S.1.2, $\mathfrak{M}$ is again a multiplication operator that depends on the spatial variable $x \in \Omega$, and $\mathfrak{K}$ is a fourth order differential operator in the spatial variable $x \in \Omega$,

$$\mathfrak{M}(\cdot) := \rho(x)A(x)(\cdot),$$

$$\mathfrak{K}(\cdot) := \frac{\partial^2}{\partial x^2}\left(C(x)I(x)\frac{\partial^2}{\partial x^2}(\cdot)\right).$$

The reader should carefully compare the similar form of the SDOF Equation 8.36 and Equation 8.49 that defines a DPS. Equation 8.36 defines a trajectory $t \mapsto u(t)$ that evolves in $\mathbb{R}$. On the other hand, Equation 8.49 defines a trajectory $t \mapsto u(t)$ that evolves in a space H of functions over the domain Ω.

Many different DPS associated with the vibration of continua have governing equations that can be put in the form of Equation 8.49. The equations governing undamped vibration of uniaxial linearly elastic rods, Bernoulli–Euler beams, Kirchoff plates and shells, and linearly elastic continua subject to suitable boundary and initial conditions can all be written in the form in Equation 8.49 [29], perhaps with some modification for the number of spatial dimensions and vector-valued state functions [4]. Some expositions generate governing equations that appear in slightly different form, but often can be shown to be equivalent or closely related to the structure in Equation 8.49. For example [4] provides a general framework for a wide variety of piezoelectric composite systems that is cast in terms of mass, damping, and stiffness bilinear forms. These bilinear forms are often derived by starting with equations of equilibrium that have a form similar to that shown in Equation 8.49.

Homogeneous Solutions: Separation of Variables

One of the most popular techniques for constructing either analytic or approximate solutions for systems as in Equation 8.49 is the method of separation of variables. We will see that this technique is particularly well suited for a DPS provided that its geometric and material properties are spatially uniform, and the geometry of the domain Ω is simple. If the DPS under consideration does not have these properties, other techniques such as Galerkin approximations should be considered.

We first describe separation of variables for the homogeneous solution of the DPS in Equation 8.43 that is obtained when we set $0 \equiv f(t) := f(t, \cdot) \in H$. The key assumption in a separation of variables approach seeks a solution of the form

$$u^h(t,x) := T(t)\psi(x),$$

with T only a function of the time t and ψ only a function of the spatial variable x. Substituting this expression into the governing Equation 8.49 and rearranging yield

$$\frac{\ddot{T}(t)}{T(t)} = -\frac{(\mathfrak{K}\psi)(x)}{(\mathfrak{M}\psi)(x)} = \lambda$$

for some unknown constant $\lambda \in \mathbb{R}$. The spatial function ψ must therefore satisfy the differential generalized eigenproblem

$$(\mathfrak{K} + \lambda\mathfrak{M})\psi = 0.$$

Note that the trivial function $\psi(x) \equiv 0$ for all $x \in \Omega$ is always a solution of this operator equation. A *nontrivial* function ψ that satisfies this equation is a generalized eigenfunction corresponding to the eigenvalue λ. Together, (λ, ψ) is known as an generalized eigenpair. If we divide through by $\mathfrak{M}$, which is a multiplication operator in our problems, and modify the definition of $\mathfrak{K}$ accordingly, we obtain

$$(\mathfrak{K} + \lambda I)\psi = 0$$

with I the identity operator. A nontrivial solution ψ and its corresponding λ constitute an eigenpair (λ, ψ) of this equation. Since $\mathfrak{K}$ is an even order differential operator, the solution of the eigenproblem requires a corresponding number of suitable boundary conditions. The boundary conditions are determined from the boundary conditions on u. The next two examples illustrate this process.

Example S.1.3 The governing equation 8.43 in Example S.1.1 with the right hand side $f(t, \cdot) := 0$ take the form

$$\rho A \frac{\partial^2 u}{\partial t^2} = \frac{\partial}{\partial x}\left(AC\frac{\partial u}{\partial x}\right),$$

subject to suitable boundary and initial conditions. To apply separation of variables, it is assumed that the cross sectional area $A := A(x)$ and Young's modulus $C := C(x)$ are constant along the length of the bar. We assume that the solution has the form $u(t, x) = T(t)\psi(x)$. It follows that we can write

$$\frac{\frac{d^2 T}{dt^2}(t)}{T(t)} = \frac{AC}{\rho A}\frac{\frac{d^2\psi}{dx^2}(x)}{\psi(x)} = \lambda,$$

for a constant $\lambda \in \mathbb{R}$. We obtain an equation for the time-varying function $T(t)$ and for the spatial function $\psi(x)$. The equation for the spatial function ψ can be written as

$$(\mathfrak{K} + \lambda\mathfrak{M})\psi = 0,$$

(Continued)

Example S.1.3 (Continued)

which is the form of the differential generalized eigenproblem. When we enforce the boundary conditions on the displacement u, we can deduce boundary conditions on ψ. Suppose that he boundary conditions are $u(t,0) = 0$ and $u(t,L) = 0$. We thereby obtain the eigenvalue problem for the spatial unknown ψ, which can be written as the ordinary differential equation

$$\frac{d^2\psi}{dx^2} = \hat{\lambda}\psi = -\omega^2\psi, \tag{8.50}$$

$$\psi(0) = 0, \tag{8.51}$$

$$\psi(L) = 0. \tag{8.52}$$

with $\hat{\lambda} := \rho\lambda/C$.

The general form of the solution is $\psi(x) := c_1 \cos\omega x + c_2 \sin\omega x$, which reduces when we apply the boundary conditions to the condition that $\psi(x) \sim \sin\omega x$ with $\omega = \frac{k\pi}{L}$ for $k \in \mathbb{N}$. Observe that the coefficient that multiplies ψ is immaterial: if ψ is and eigenfunction, then so is $\alpha\psi$ for any scalar α. We say that the eigenfuncions are determined only up to a scalar multiple. The collection of all such solution pairs $\{\hat{\lambda}_k, \psi_k\}_{k\in\mathbb{N}} = \{-\omega_k^2, \psi_k\}_{k\in\mathbb{N}} \equiv \left\{-\left(\frac{k\pi}{L}\right)^2, \sin\frac{k\pi x}{L}\right\}_{k\in\mathbb{N}}$ are the eigenvalues and eigenfunctions, or eigenpairs, of the vibration problem. If we repeat the above analysis, but choose the boundary conditions $u(t,0) = 0$

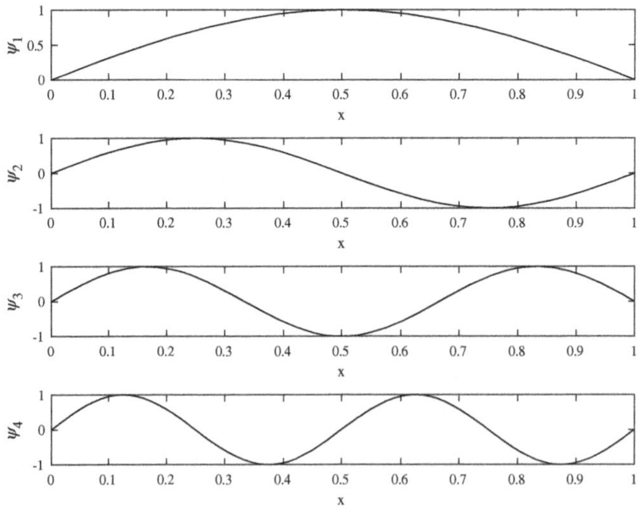

Figure 8.16 First four fixed-fixed modes ψ_i for the axial element

and $\frac{\partial u}{\partial x}(t,L) = 0$, we find that the set of eigenpairs are $\{\hat{\lambda}_k, \psi_k\}_{k\in\mathbb{N}} = \{-\omega_k^2, \psi_k\}_{k\in\mathbb{N}} \equiv \left\{ -\left(\left(k-\frac{1}{2}\right)\frac{\pi}{L}\right)^2, \sin\frac{(k-\frac{1}{2})\pi x}{L} \right\}_{k\in\mathbb{N}}$.

It can be shown from the spectral theory for compact, self-adjoint operators on a Hilbert space as discussed in Appendix S.1.2, or by direct computation, that the eigenvectors corresponding to distinct eigenvalues are orthogonal in $L^2(0,L)$. In particular, we have

$$(\psi_m, \psi_n)_{L^2(\Omega)} := \int_0^L \sin\frac{m\pi x}{L}\sin\frac{n\pi x}{L}dx = \begin{cases} \frac{L}{2} & \text{if } m = n \\ 0 & \text{otherwise,} \end{cases}$$

for either case of boundary conditions. As noted earlier the eigenfunctions ψ_k are defined only up to a scalar: ψ_k is an eigenfunction if and only if $\alpha\psi_k$ is an eigenfunction for any $\alpha \in \mathbb{R}$. A particularly useful scaling chooses

$$\psi_k(x) := \frac{\sin\frac{k\pi x}{L}}{\sqrt{\int_0^L \sin^2\frac{k\pi\eta}{L}d\eta}}.$$

With this choice of scaling, the eigenfunctions ψ_k are orthonormal in the sense that $(\psi_i, \psi_j) = \delta_{i,j}$. Unless otherwise noted, we will use this normalization in the definition of eigenfunctions.

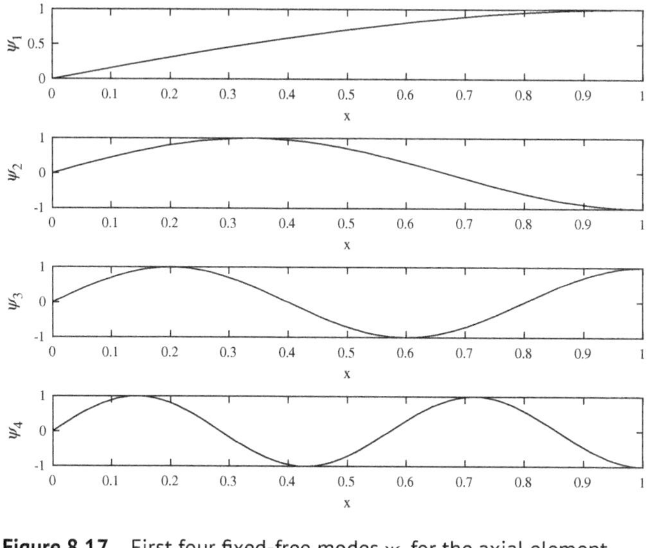

Figure 8.17 First four fixed-free modes ψ_i for the axial element

(Continued)

Example S.1.3 (Continued)

Since the governing Equation 8.43 is linear, we obtain the following expression for the homogeneous solution

$$u^h(t,x) := \sum_{k=1}^{\infty} T_k(t)\psi_k(x)$$

by superposition. Substituting this equation into the governing Equation 8.43 yields

$$\sum_{k=1}^{n} \{\rho A \ddot{T}_k(t) + AC\omega_k^2 T_k(t)\}\psi_k(x) = 0, \tag{8.53}$$

when we use the eigenvalue equation $-\psi_k'' = \omega_k^2 \psi_k$ for $k \in \mathbb{N}$. From orthogonality of the modes ψ_k, it follows that

$$\ddot{T}_k(t) + \frac{C}{\rho}\omega_k^2 T_k(t) = 0.$$

Referring to Section S.1.1, the general form of the solution of the homogeneous equation is

$$u^h(t,x) = \sum_{k=1}^{\infty} (A_k \cos \omega_{n,k} t + B_k \sin \omega_{n,k} t) \frac{\sin\left(\frac{k\pi x}{L}\right)}{\sqrt{\int_0^L \sin^2 \frac{k\pi\eta}{L} d\eta}}$$

with the natural frequency $\omega_{n,k} := \sqrt{C/\rho}\omega_k$.

Example S.1.4 The governing equation for the beam in Example S.1.2 takes the form

$$\rho A \frac{\partial^2 u}{\partial t^2} = -\frac{\partial^2}{\partial x^2}\left(CI\frac{\partial^2 u}{\partial x^2}\right), \tag{8.54}$$

which follows from Equation 7.23 by letting the piezoelectric coupling constant $e_{31} = 0$. We will solve this equation for the most common boundary conditions, those that model a cantilever beam and are summarized in the table in Example 8.3.2. The beam is cantilevered at $x = 0$ so that

$$u(t,0) = 0,$$

$$\frac{\partial u}{\partial x}(t,0) = 0.$$

The beam is free at $x = L$, which means that the shear force S and the bending moment $\mathcal{M}$ are equal to zero at $x = L$. We have

$$S(t, L) = -\frac{\partial}{\partial x}\left(CI\frac{\partial^2 u}{\partial x^2}\right)\bigg|_{x=L} = -CI\frac{\partial^3 u}{\partial x^3}(t, L) = 0, \tag{8.55}$$

$$\mathcal{M}(t, L) = -CI\frac{\partial^2 u}{\partial x^2}(t, L) = 0. \tag{8.56}$$

Note that here we have assumed that the material constants ρ, A, C, I are uniform along the beam. We choose to represent the displacement $u(t, x)$ in the separable form

$$u(t, x) = T(t)\psi(x). \tag{8.57}$$

When Equation 8.57 is substituted into Equation 8.54, two independent equations for the time and spatial functions are produced from

$$\frac{\frac{d^2 T}{dt^2}(t)}{T(t)} = -\frac{CI}{\rho A}\frac{\frac{d^4\psi}{dx^4}(x)}{\psi(x)} = \lambda. \tag{8.58}$$

The equation implied above for the spatial function ψ can again be written in the form of a generalized eigenproblem

$$(\mathfrak{K} + \lambda\mathfrak{M})\psi = 0$$

for some $\lambda \in \mathbb{R}$. The eigenproblem for the spatial variable ψ can then be written as

$$\frac{d^4\psi}{dx_1^4} = \omega^4\psi, \tag{8.59}$$

with $w^4 := -\frac{\rho A}{CI}\lambda$, The boundary conditions on $u(t, x)$ imply that the boundary conditions on the eigenproblem are

$$\psi(0) = 0, \qquad \frac{d^2\psi}{dx_1^2}(L) = 0, \tag{8.60}$$

$$\frac{d\psi}{dx_1}(0) = 0, \qquad \frac{d^3\psi}{dx_1^3}(L) = 0. \tag{8.61}$$

The details of solving the eigenproblem for a fourth order differential operator are lengthy and will not be repeated here. The solution procedure is presented in standard textbooks on vibrations [21] or on structural dynamics [11].

(Continued)

Example S.1.4 (Continued)

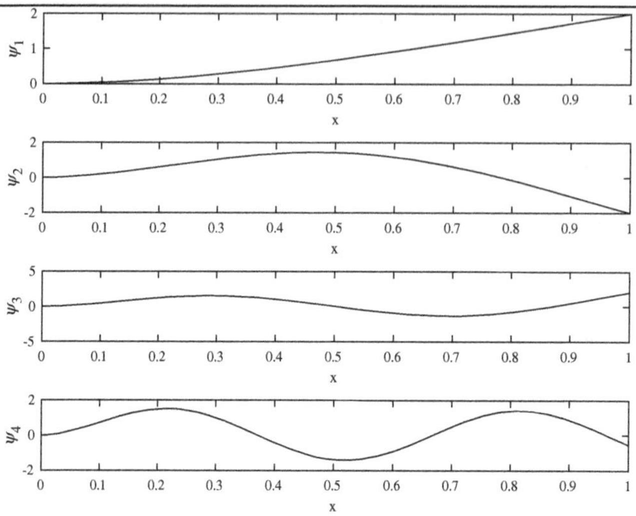

Figure 8.18 First four cantilever modes ψ_i of the Bernoulli–Euler beam

The kth eigenfunction of Equation 8.59 is proportional to

$$\psi_k(x) := \cosh \omega_k x - \cos \omega_k x - \alpha_k(\sinh \omega_k x - \sin \omega_k x),$$

where ω_k is the kth solution of the transcendental equation

$$\cos \omega_k L \cosh \omega_k L = -1,$$

and

$$\alpha_k = -\frac{\cosh \omega_k L + \cos \omega_k L}{\sinh \omega_k L + \sin \omega_k L}.$$

As in our study of linearly elastic axial rods, the modes ψ_k are orthogonal in the sense that

$$\int_0^L \psi_i(\eta)\psi_j(\eta)d\eta = \begin{cases} \int_0^L \psi_i^2(\eta)d\eta & \text{if } i = j, \\ 0 & \text{otherwise.} \end{cases}$$

This orthogonality property can be demonstrated by direct integration or by using the spectral theory summarized in S.1.2 for compact, self-adjoint operators on a Hilbert space. As in the last example, we choose the normalization of ψ_k so that $\int_0^L \psi_k(\eta)\psi_m(\eta)d\eta = \delta_{k,m}$. The coefficients ω_k and α_k can be estimated numerically, and their approximations tabulated below can be found in [21].

mode number k	$\omega_k L$	α_k
1	1.87510407	0.7341
2	4.69409113	1.0185
3	7.85475744	0.9992
4	10.59554073	1.0000
5	14.13716839	1.0000
$k > 5$	$\frac{(2k-1)\pi}{2}$	1

We again construct the superposition of solutions that have a separable form

$$u(t,x) := \sum_{k=1}^{\infty} T_k(t)\psi_k(x),$$

and substitute it into Equation 8.54

$$\sum_{k=1}^{n} \{\rho A \ddot{T}_k(t) + CI\omega_k^4 T_k(t)\}\psi_k(x) = 0, \tag{8.62}$$

From the orthogonality of the modes ψ_k, it follows that

$$T_k(t) = A_m \cos \mu\omega_k^2 t + B_k \sin \mu\omega_k^2 t.$$

The homogeneous solution is then

$$u^h(t,x) = \sum_{k=1^\infty} (A_m \cos \omega_{n,k}^2 t + B_k \sin \omega_{n,k}^2 t)\psi_k(x).$$

with the natural frequency $\omega_{n,k} := \sqrt{CI/\rho A}\omega_k^2$.

If we are only interested in the homogeneous solution, so that the particular solution $u^p(t,\cdot) \equiv 0$, the unknown constants $\{(A_k, B_k)\}_{k=1}^{\infty}$ in Examples S.1.1 and S.1.2 can be determined from the initial conditions on $T_k(0)$ and $\dot{T}_k(0)$. We show how this is done after Section S.1.2 which provides a rigorous justification for the process. The theory presented in Section S.1.2 is rather advanced, and the reader can omit this section in a first reading.

Eigenvalue Problems in Hilbert Spaces

In our discussion of separation of variable methods for DPS, we have relied on the determination of eigenfunctions $\{\psi_k\}_{k=1}^{\infty}$. We have noted that these functions have remarkable properties that facilitate the solution of the governing equations of some DPS. In particular we have scaled the eigenfunctions so that they are orthonormal in the sense that $(\psi_k, \psi_\ell) = \delta_{k,\ell}$. While the orthonormality of the eigenfunctions can be verified case by case by direct integration, *spectral theory* on Hilbert spaces H provides a guarantee of this property for many problems in mechanical vibrations. It also establishes that the eigenfunctions can be used to approximate functions in H with an arbitrarily small error.

Our discussion begins with a few basic definitions. An inner product space $(H, (\cdot,\cdot))$ over the complex numbers $\mathbb{C}$ is a vector space H that is equipped with an inner product $(\cdot,\cdot)$ [33]. The inner product $(\cdot,\cdot) : H \times H \to \mathbb{C}$ is a function that satisfies the following properties:

H1) $(u, u) \geq 0$ and equality holds only when $u = 0$,
H2) $(u, v) = \overline{(v, u)}$ for all $u, v \in H$,
H3) $(\alpha u + \beta v, w) = \alpha(u, w) + \beta(v, w)$ for all $u, v, w \in H$ and $\alpha, \beta \in \mathbb{C}$.

An inner product space $(H, (\cdot,\cdot))$ over the real numbers $\mathbb{R}$ is defined similarly. In either case an inner product space can always be equipped with a norm by defining $\| f \| := \sqrt{(f,f)}$, and if the inner product space is complete in this norm, it is said to be a Hilbert space. For the Fourier approximations discussed in this text the most important example of a Hilbert space is the space $L^2(D)$ of real-valued, square integrable functions $f : \Omega \to \mathbb{R}$ over the domain Ω. The inner product on this space of functions is defined as

$$(f, g) := \int_\Omega f(\eta)g(\eta)d\eta.$$

For example, when we consider uniaxial rods or beams of length L, an appropriate Hilbert space for functions that represent deformation will be $L^2(0, L)$. If we consider square plates having dimension $L \times L$, we may choose to represent lateral deformation as a function that lives in the Hilbert space $L^2((0, L) \times (0, L))$.

An operator $L : H \to H$ is a linear operator if for any scalars α, β we have $L(\alpha u + \beta v) = \alpha Lu + \beta Lv$ for all $u, v \in H$, and it is bounded if the operator norm $\| L \|$ of L is bounded in the definition

$$\| L \| := \sup_{u \neq 0} \frac{\| Lu \|}{\| u \|} < \infty.$$

A bounded, linear operator L is self-adjoint if $(Lu, v) = (u, Lv)$ for all $u, v \in H$. A linear operator is compact if it maps any bounded sequence into one that contains a convergent subsequence [24].

An eigenvector ψ of the operator equation

$$L\psi = \lambda\psi$$

is any nontrivial solution, that is, any solution such that $\psi \neq 0$. The constant λ is the eigenvalue associated with the eigenvector ψ. The eigenspace associated with the eigenvalue λ is defined to be span$\{\psi : L\psi = \lambda\psi\}$. The following theorem is sometimes referred to as the spectral theorem for compact, self-adjoint, linear operators. Various forms of this theorem can be found, for example, in [24, 33].

Theorem S.1.1 Let $L : H \to H$ be a linear, bounded, self-adjoint, compact operator on the complex Hilbert space H. Then the following hold:

1. The eigenvalues of L are real.
2. The eigenvectors of L corresponding to distinct eigenvalues are orthogonal.
3. Each eigenspace of L is finite dimensional.
4. For each positive constant $c > 0$, the number of eigenvalues λ of L that are greater than c, $\lambda > c$, is finite.
5. There exists an orthonormal basis $\left\{\psi_k\right\}_{k\in\mathbb{N}}$ of H that consists of eigenvectors of L. In other words, $L\psi_k = \lambda_k\psi_k$ for each $k \in \mathbb{N}$, and the set $\{\psi_k\}_{k\in\mathbb{N}}$ is linearly independent and dense in H.

Proof: The proof of various parts of this theorem can be found in many places: see [33] or [24] for a detailed discussion. Some of the conclusions can be derived quite expediently. Suppose that λ_k, ψ_k and λ_j, ψ_j are distinct eigenpairs. We then have

$$
\begin{aligned}
L\psi_j &= \lambda_j\psi_j, \\
L\psi_k &= \lambda_k\psi_k.
\end{aligned}
$$

If we take the inner product of the first equation above with ψ_k, take the complex conjugate of the resulting expression, take the inner product of the second with ψ_j, and subtract the results, we find that

$$
0 = (\overline{\lambda}_j - \lambda_k)\overline{(\psi_j, \psi_k)}.
$$

We conclude that $\mathrm{im}(\lambda_j) = 0$ when we choose $j = k$ in the above equation. When $j \neq k$, it follows that ψ_j and ψ_k are orthogonal. When we say that the set of eigenvectors $\{\psi_k\}_{k\in\mathbb{N}}$ is dense in H, this means that for any $f \in H$, we can find a set of scalars $\{\alpha_k\}_{k\in\mathbb{N}}$ such that

$$
f = \lim_{N\to\infty} \sum_{k=1}^{N} \alpha_k\psi_k.
$$

In other words, the eigenvectors can be used to build a norm-convergent approximation of any $f \in H$.

This powerful theorem is used to justify many modal approximation strategies in vibrations and structural dynamics.

Example S.1.5 In Example 8.3.3 we solved the eigenproblem

$$\frac{d^2\psi}{dx^2} = \lambda\psi = -\omega^2\psi,$$
$$\psi(0) = 0,$$
$$\psi(L) = 0.$$

The collection of all eigenpairs is the countable collection

$$\{(\lambda_k, \psi_k)\}_{k=1,\dots,\infty} = \left\{\left(-\left(\frac{k\pi x}{L}\right)^2, \sin\frac{k\pi x}{L}\right)\right\}_{k=1,\dots,\infty}$$

Examples 8.3.3 and S.1.3 note without proof that the eigenfunctions corresponding to distinct eigenvalues are orthogonal. This property of the eigenvectors $\{\psi_k\}_{k\in\mathbb{N}}$ of the operator $L(\cdot) = \frac{d^2}{dx^2}(\cdot)$ can be deduced using Theorem S.1.1. It is well documented [32] that the inverse operator L^{-1} of the operator L is an integral operator that is bounded, linear, self-adjoint, and compact. Furthermore, since

$$L^{-1}\psi_k = \frac{1}{\lambda_k}\psi_k \qquad \Leftrightarrow \qquad L\psi_i = \lambda_k\psi_k,$$

we know that ψ_k is an eigenvector of L associated with the eigenvalue λ_k if and only ψ_k is an eigenvector of L^{-1} associated with the eigenvalue $\frac{1}{\lambda_k}$. By applying Theorem S.1.1 to the integral operator L^{-1}, we conclude that distinct eigenvectors are orthogonal. Example 8.3.3 uses the eigenmodes ψ_k to represent an unknown displacement field. It might be questioned as to why this is a reasonable representation. Again, this choice is justified by the spectral theorem S.1.1. The eigenmodes are a basis: they are linearly independent and dense in $H = L^2(0, L)$. It is for this reason that any function $u \in C([0, \infty), H)$ has an expansion such as

$$u(t, x) = \sum_k \alpha_k(t)\psi_k(x)$$

such that the spatial function $u(t, \cdot)$ converges in $L^2(0, L)$ at each $t > 0$.

Particular Solutions and FRFs

Examples S.1.3 and S.1.4 present techniques for solving for the homogeneous solutions $u^h(t, \cdot)$ for the partial differential equations in Examples 8.43 and 8.46. In this section we present methods that can be applied to solve for the frequency response function (FRF) for such systems.

We consider again the evolutionary PDEs in Equation 8.49 and seek a $u(t) \in H$ that solves

$$\mathfrak{M}\frac{\partial^2 u}{\partial t^2}(t) + \mathfrak{C}\dot{u}(t) + \mathfrak{K}u(t) = f(t) \tag{8.63}$$

subject to suitable boundary conditions and initial conditions, where $\mathfrak{M}$ is a zero order differential operator, $\mathfrak{K}$ is an even order spatial differential operator, and $\mathfrak{C}$ is a differential operator representing damping or energy dissipation. We make the following assumptions in what follows:

A1) The eigenpairs $\{(\lambda_k, \psi_k)\}_{k=1}^{\infty}$ are solutions of the differential generalized differential eigenproblem $(\mathfrak{K} + \lambda_k \mathfrak{M})\psi_k = 0$ for appropriate boundary conditions associated with Equation 8.63. The eigenfunctions satisfy the $\mathfrak{M}$-orthonality and $\mathfrak{K}$-orthogonality conditions $(\mathfrak{M}\psi_\ell, \psi_j)_H = m_\ell \delta_{\ell j}$ and $(\mathfrak{K}\psi_\ell, \psi_j)_H = k_\ell \delta_{\ell j}$, respectively.
A2) The damping operator $\mathfrak{C}$ satisfies $(\mathfrak{C}\psi_\ell, \psi_j) = c_\ell \delta_{\ell j}$ for the generalized eigenfunctions $\{\psi_k\}_{k=1}^{\infty}$ of $\mathfrak{M}$ and $\mathfrak{K}$ in Assumption (A1).
A3) The function $f(t) \in H$ can be written as $f(t) := \mathfrak{M}p(t)$ for some function $p(t)$ for all $t \in \mathbb{R}^+$ and $p(t) \in H$ can be written in the form

$$\sum_{k=1\infty} P_k(t)\psi_k(x).$$

It should be noted that the above assumption (A1) implies that the eigenvalues satisfy $k_j + \lambda_j m_j = 0$, or that $\lambda_j := -k_j/m_j$ for $j \in \mathbb{N}$.

We represent the steady state solution in the separation of variables form

$$u^p(t,x) = \sum_k T_k(t)\psi_k(x),$$

substitute this expression into Equation 8.63, and take the inner product with an arbitrary function ψ_ℓ to obtain

$$\begin{aligned} 0 &= \{(\psi_\ell, \mathfrak{M}\psi_k)\ddot{T}_k(t) + (\psi_\ell, \mathfrak{C}\psi_k)\dot{T}_k(t) + (\psi_\ell, \mathfrak{K}\psi_k)_k(t) - (\psi_\ell, \mathfrak{M}\psi_k)P_k(t)\}, \\ &= \{m_\ell \delta_{\ell,k}\ddot{T}_k(t) + c_\ell \delta_{\ell,k}\dot{T}_k(t) + k_\ell \delta_{\ell,k}T_k(t) - m_\ell \delta_{\ell,k}P_k(t)\}, \end{aligned}$$

with the summation convention over $k = 1, \ldots, \infty$. We must have

$$\ddot{T}_\ell(t) + 2\xi_\ell \omega_{n,\ell}\dot{T}(t) + \omega_{n,\ell}^2 T_\ell = P_\ell(t)$$

for all $t \in \mathbb{R}^+$ and $\ell = 1, \ldots, \infty$. But the above equations are decoupled, SDOF equations of motion as in Section S.1.1, so we define $2\xi_\ell \omega_{n,\ell} = c_\ell/m_\ell$ and $\omega_{n,\ell}^2 = k_\ell/m_\ell$. We then assume that $P_k(t) := P_{0,k}e^{\hat{j}\varpi t}$ and $T_k(t) := T_k e^{\hat{j}\varpi t}$ with $P_{0,k}$ the real amplitude of the excitation for mode k and T_k the complex response of mode k, respectively. Following the same steps described in Section S.1.1 for SDOF systems, we obtain the FRF $H_k(\hat{j}\varpi)$ as

$$H_k(\hat{j}\varpi) := \frac{1}{(1 - r_k^2) + 2\xi_k r_k \hat{j}}\frac{1}{\omega_{n,k}^2}$$

with $r_k := \varpi/\omega_{n,k}$. The complex response in terms of the FRF is then

$$u(t,x) := \sum_{k=1}^{\infty} \frac{1}{\sqrt{(1-r_k^2)^2 + (2\xi_k r_k)^2}} \frac{1}{\omega_{n,k}^2} P_{0,k} e^{(\hat{j}\varpi t + \beta_k)} \psi_k(x),$$

with the phase angle given as

$$\beta_k = -\arctan \frac{2\xi_k r_k}{1 - r_k^2}.$$

S.1.3 MDOF Equations of Motion

Generalized Algebraic Eigenvalue Problems

Appendix S.1.2 summarizes the theory of eigenvalue problems for self adjoint, compact operators on a Hilbert space, and the use of their eigenfunctions as the basis for Galerkin approximations is discussed in Section 8.3. The resulting Fourier or modal approach has been shown to be advantageous in that it diagonalizes certain classes of evolutionary partial differential equations.

When using finite element shape functions as the basis of a Galerkin approximation, in contrast, the resulting equations are perhaps sparse and banded, but not necessarily diagonal. In such cases the generalized algebraic eigenvalue problem can be used to construct decoupled equations. A nontrival vector $v \in \mathbb{R}^n$ is a generalized eigenvector of the matrices $A, B \in \mathbb{R}^{n \times n}$ corresponding to the generalized eigenvalue λ if it satisfies the equation

$$Av = \lambda Bv.$$

As in the definition of the eigenvector in Section S.1.2, it is essential that v is a nontrivial solution, $v \not\equiv 0$. Similar to the orthogonality of eigenfunctions described in the spectral theorem in Theorem S.1.1, the eigenvectors of the generalized algebraic eigenproblem exhibit important orthogonality properties that have widespread use in vibrations applications.

Theorem S.1.2 Suppose that $A, B \in \mathbb{C}^{n \times n}$ are Hermitian matrices, $\overline{A}^T = A, \overline{B}^T = B$, and B is positive definite. Then

1. Any generalized eigenvalue λ_i is real.
2. If v_i and v_j are generalized eigenvectors of A, B corresponding to distinct eigenvalues $\lambda_i \neq \lambda_j$, they are A-orthogonal and B-orthogonal in the sense that

$$v_i^T A v_j = \begin{cases} 0 & \text{if } i \neq j, \\ a_i = v_i^T A v_i & \text{if } i \equiv j, \end{cases}$$

$$v_i^T B v_j = \begin{cases} 0 & \text{if } i \neq j, \\ b_i = v_i^T B v_i & \text{if } i \equiv j. \end{cases}$$

Proof: Suppose that v_i, v_j are generalized eigenvectors corresponding to eigenvalues λ_i and λ_j, respectively. When we take the inner product of the eigenproblem for v_i with v_j, and take the inner product of the eigenproblem for v_j with v_i, we find

$$(Av_i, v_j) = \lambda_i(Bv_i, v_j), \tag{8.64}$$

$$(Av_j, v_i) = \lambda_j(Bv_j, v_i). \tag{8.65}$$

The second Equation 8.65 can be expressed as

$$\overline{(Av_i, v_j)} = \lambda_j\overline{(Bv_i, v_j)}.$$

Subtracting this equation from the conjugate of Equation 8.64, and using the conjugate symmetry of A, B implies the $(\overline{\lambda}_i - \lambda_j)\overline{(Bv_i, v_j)} = 0$. If we choose $i \equiv j$, the positive definiteness of B implies that $(Bv_i, v_i) > 0$ and that $\overline{\lambda}_i = \lambda_i$. In other words, λ_i is real. If on the other hand, $\lambda_i \neq \lambda_j$, we conclude that v_i and v_j are B-orthogonal. The A-orthogonality of the two eigenvectors then follows from either of the Equations 8.64 or 8.65 and the B-orthogonality of the eigenvectors.

Example S.1.6 In this text we use Theorem S.1.2 to decouple equations of motion. Suppose that we have derived the equations of motion that result from using finite element shape functions as the basis in a Galerkin approximation, $u(t,x) := N_j(x)u_j(t)$,

$$m_{ij}\ddot{u}_j(t) + k_{ij}u_j(t) = b_iV(t). \tag{8.66}$$

Let $\phi_{j,k} \in \mathbb{R}^n$ be the jth entry of the kth generalized eigenvector of the generalized algebraic eigenvalue problem,

$$k_{ij}\phi_{j,k} = \lambda_k m_{ij}\phi_{j,k}. \tag{8.67}$$

where the summation convention runs over $j, k = 1, \dots n$. We assume that the degrees of freedom u_j can be represented in terms of a linear combination of the generalized eigenvectors

$$u_j(t) := \phi_{j,\ell}\eta_\ell(t). \tag{8.68}$$

The coefficients $\eta_k(t)$ are called the generalized modal degrees of freedom for $\ell = 1, \dots, n$. The modal expansion in Equation 8.68 is substituted into the governing Equation 8.67. We derive the uncoupled equations when we multiply the result by $\phi_{\ell,i}$

$$m_\ell\ddot{\eta}(t) + k_\ell\eta(t) = \phi_{i,\ell}b_iV(t),$$

with the summation only over $i = 1, \dots, n$. The equations are written in terms of the modal mass $m_\ell := \phi_{i,\ell}m_{ij}\phi_{j,\ell}$ and the modal stiffness $k_\ell := \phi_{i,\ell}k_{ij}\phi_{j,\ell}$.

(Continued)

Example S.1.6 (Continued)

These equations for the modal degrees of freedom are uncoupled, and any of the solution methods reviewed in Appendix in Section S.1.1 can be applied. Suppose that we are interested in determining the steady state response of the modal coordinates η_ℓ for $\ell = 1, \dots, n$ for a harmonic input $V(t) = \frac{1}{m_\ell} b_\ell V_0 \cos \varpi t$. Let us also assume that the modal equations are subject to viscous damping defined in terms of a damping matrix c_{ij} that happens to be diagonalized by the generalized eigenvectors $\phi_{i,\ell}$

$$\phi_{i,\ell} c_{ij} \phi_{j,k} = \begin{cases} 0 & \text{if } \ell \neq k, \\ c_\ell & \text{if } \ell = k. \end{cases}$$

Carefully note that the generalized eigenvectors are not guaranteed by construction to diagonalize the damping matrix. They are derived from the mass and stiffness matrices and are only guaranteed to be mass and stiffness orthogonal. This additional assumption about diagonalizing the viscous damping matrix is quite commonly made in engineering applications. The steady state response of the modal coordinate $\eta_\ell(t)$ is given by the expression

$$\eta_\ell(t) = |H_\ell(\hat{j}\varpi)| \frac{1}{m_\ell} b_\ell V_0 \cos(\varpi t + \beta_\ell).$$

The summation convention is not used in this equation; it is a single equation indexed by ℓ. In this equation the frequency response function $H_\ell(\hat{j}\varpi)$ is determined by its magnitude

$$|H_\ell(\hat{j}\varpi)| = \frac{1}{\sqrt{(1 - r_\ell^2)^2 + (2\xi_\ell r_\ell)^2}} \frac{1}{\omega_{n,\ell}^2},$$

and phase $\beta_\ell = -\arctan \frac{2\xi_\ell r_\ell}{1 - r_\ell^2}$, in which $\omega_{n,\ell}^2 = \frac{k_\ell}{m_\ell}$, $2\xi_\ell \omega_{n,\ell} = \frac{c_\ell}{m_\ell}$, and $r_\ell = \frac{\varpi}{\omega_{n,\ell}}$. The physical response or displacement is obtained by summing the contributions of all the modal degrees of freedom $\eta_\ell(t)$ for $\ell = 1, \dots, n$ in the Galerkin approximation $u(t,x) = N_j(x) u_j(t) = N_j(x) \phi_{j,\ell} \eta_\ell(t)$. The final result becomes

$$u(t,x) = N_j(x) \phi_{j,\ell} |H_\ell(\hat{j}\varpi)| \frac{1}{m_\ell} b_\ell V_0 \cos(\varpi t + \beta_\ell)$$

with the summation over $j, \ell = 1, \dots, n$.

S.2 Tensor Analysis

Chapter 2 provides a succinct and abbreviated account of tensor analysis as it is applied in this text. In this appendix we summarize a more general treatment of tensor analysis. The goal is to enable those having the requisite background to see how the special case studied in this text fits in the general theory, and to illustrate to beginning students how the theory is extended to the general case. An operator $T : U \to V$ that acts between the real vector spaces U and V is linear if we have $T(\alpha u_1 + \beta u_2) = \alpha T(u_1) + \beta T(u_2)$ for all $u_1, u_2 \in U$ and all $\alpha, \beta \in \mathbb{R}$. We denote by $L(U, V)$ the vector space of all linear operators from U into V,

$$L(U, V) := \left\{ T : U \to V \mid T \text{ is linear} \right\}.$$

When the dimension $\dim(U)$ of U and $\dim(V)$ of V are finite, we always have $\dim(L(U, V)) = \dim(U) \cdot \dim(V)$. If we are given a finite collection of real vector spaces U_i for $i = 1, \dots, n$, a mapping $T : U_1 \times U_2 \times \cdots \times U_n \to V$ from the vector space $U_1 \times \cdots \times U_n$ into the vector space V is multilinear if it is linear in each of its arguments. In this case for each $k = 1, \dots, n$ we have

$$T(u_1, \dots, \alpha u_k + \beta w_k, \dots, u_n) = \alpha T(u_1, \dots, u_k, \dots, u_n) + \beta T(u_1, \dots, w_k, \dots, u_n).$$

for all $u_i \in U_i$ and $i = 1, \dots n$, $w_k \in U_k$, and $\alpha, \beta \in \mathbb{R}$.

Definition S.2.1 Let U be a real vector space. The collection of tensors $T_s^r(U)$ of type $(r, s) \in \mathbb{N} \times \mathbb{N}$ is the vector space that consists of multilinear maps taking values in $\mathbb{R}$,

$$T_s^r(U) = L(\underbrace{U^* \times \ldots \times U^*}_{r \text{ products}}, \underbrace{U \times \ldots \times U}_{s \text{ products}}, \mathbb{R}).$$

In other words, a tensor of order (r, s) is a real-valued, multilinear map that takes r arguments from the dual space U^* and s arguments from the vector space U. While we consider only orthonormal bases in this text, as emphasized in Chapter 2, in general a basis of a vector space U is a linearly independent set of vectors that span U. Such a collection of vectors need not be orthogonal, nor have unit length. If $\{g_i\}_{i=1}^n$ is such a basis for U, a fundamental theorem [9, 33] guarantees that there is a unique dual basis $\{g^j\}_{j=1}^n$ of U^* that is biorthogonal in the sense that

$$\langle g^j, g_i \rangle_{U^* \times U} := g^j(g_i) = \delta_i^j.$$

The notation $\langle \cdot, \cdot \rangle_{U^* \times U} : U^* \times U \to \mathbb{R}$ is referred to as the duality pairing on $U^* \times U$. Since U^* is itself a vector space, it is possible to consider also the duality pairing on $U^{**} \times U^*$ denoted by $\langle \cdot, \cdot \rangle_{U^{**} \times U^*}$. However, this notation is seldom seen when treating finite dimensional vector spaces. If we have a particular $u \in U$ fixed, we can always define an associated element $u^{**} \in U^{**}$ in terms of an element $u \in U$ by setting

$$u^{**}(u^*) := u^*(u).$$

for all $u^* \in U^*$. The above association of $u \in U$ with an element $u^{**} \in U^{**}$ defines an injective mapping or embedding $\mathcal{I} : U \to U^{**}$. This embedding $\mathcal{I}$ is sometimes called the canonical injection $\mathcal{I}$ of U into U^{**}. Since we deal only with finite dimensional vector spaces, the canonical injection of $U \to U^{**}$ is onto: in other words, U is isomorphic to U^{**} [33]. This means that that we can write

$$\langle \mathcal{I} g_i, g^j \rangle_{U^{**} \times U^*} = \langle g_i, g^j \rangle_{U \times U^*} := g^j(g_i).$$

It is conventional then to dispense with the explicit reference to the embedding $\mathcal{I}$ and just write $\langle g_i, g^j \rangle_{V \times V^*}$ for the duality pairing on $V^{**} \times V^*$. In fact, it is also commonplace to suppress the explicit subscript in a duality pairing such as $\langle \cdot, \cdot \rangle_{U^* \times U}$ or $\langle \cdot, \cdot \rangle_{U \times U^*}$ and just write $\langle \cdot, \cdot \rangle$. The appropriate subscript is determined from the arguments or by context.

With this interpretation of the duality pairing $\langle \cdot, \cdot \rangle$ in mind, the collection of multilinear, real valued operators

$$\{g_{i_1} \otimes \cdots \otimes g_{i_r} \otimes g^{j_1} \otimes \cdots \otimes g^{j_s} |\ i_1, \ldots, i_r, j_1 \ldots j_s \in \{1, \ldots n\}\}$$

are a basis for $T^r_s(U)$. If $T \in T^r_s(U)$, then it has a unique expansion

$$T = T^{i_1 \cdots i_r}_{j_1 \cdots j_s} g_{i_1} \otimes \cdots \otimes g_{i_r} \otimes g^{j_1} \otimes \cdots \otimes g^{j_s}.$$

Each of the components $T^{i_1 \cdots i_r}_{j_1 \cdots j_s}$ can be computed directly using the biorthogonality of the bases $\{g_i\}_{i=1}^n$ and $\{g^j\}_{j=1}^n$, as demonstrated in the following example.

Example S.2.1 Since $T \in T^r_s(U)$, it is a multilinear operator on $(\otimes U^*)^r \otimes (\otimes U)^s$. We choose to evaluate T on an arbitrary basis element $g^{k_1} \otimes \cdots \otimes g^{k_s} \otimes g_{m_1} \otimes \cdots \otimes g_{m_s}$ of this space. We compute

$$\begin{aligned} T(g^{k_1} &\otimes \cdots \otimes g^{k_s} \otimes g_{m_1} \otimes \cdots \otimes g_{m_s}) \\ &= T^{i_1 \cdots i_r}_{j_1 \cdots j_s} g_{i_1} \otimes \cdots \otimes g_{i_r} \otimes g^{j_1} \otimes \cdots \otimes g^{j_s}(g^{k_1} \otimes \cdots \otimes g^{k_s} \otimes g_{m_1} \otimes \cdots \otimes g_{m_s}), \\ &= T^{i_1 \cdots i_r}_{j_1 \cdots j_s} < g_{i_1}, g^{k_1} >_{U \times U^*} + \cdots + < g^{j_s}, g_{m_s} >_{U^* \times U} \\ &= T^{i_1 \cdots i_r}_{j_1 \cdots j_s} \delta^{k_1}_{i_1} \cdots \delta^{j_s}_{m_s} \\ &= T^{k_1 \cdots k_r}_{m_1 \cdots m_s}. \end{aligned}$$

In other words, we have shown that

$$T^{k_1 \cdots k_r}_{m_1 \cdots m_s} = T(g^{k_1} \otimes \cdots \otimes g^{k_s} \otimes g_{m_1} \otimes \cdots \otimes g_{m_s}).$$

In Chapter 1 emphasis was given on the methods for transforming the components of a tensor relative to one basis into the components relative to a different basis. We consider a second order tensor as an exemplar in the discussion that follows: higher order tensors follow the same general pattern, but the notation can become tedious. Suppose that $T \in L(U^* \times U, \mathbb{R})$ is a second order tensor of type $T^1_1(U)$. From the discussions above we know that we have a representation

$$T = T^j_i g_j \otimes g^i. \tag{8.69}$$

If we are given another pair of dual bases $\{\tilde{g}^i\}_{i=1}^n$ and $\{\tilde{g}_j\}_{j=1}^n$, it is also possible to express T in the form

$$T = \tilde{T}^j_i \tilde{g}_j \otimes \tilde{g}^i. \tag{8.70}$$

When we calculate the action of an arbitrary basis element $\tilde{g}^k \otimes \tilde{g}_m$ on Equations 8.69 and 8.70, and set the resulting terms equal, we obtain

$$\langle \tilde{g}^k \otimes \tilde{g}_m, \tilde{T}^j_i \tilde{g}_j \otimes \tilde{g}^i \rangle = \langle \tilde{g}^k \otimes \tilde{g}_m, T^j_i g_j \otimes g^i \rangle,$$

$$\langle \tilde{g}^k, \tilde{g}_j \rangle \langle \tilde{g}_m, \tilde{g}^i \rangle \tilde{T}^j_i = \langle \tilde{g}^k, g_j \rangle \langle \tilde{g}_m, g^i \rangle T^j_i \tag{8.71}$$

$$\delta^k_j \delta^i_m \tilde{T}^j_i = \langle \tilde{g}^k, g_j \rangle \langle \tilde{g}_m, g^i \rangle T^j_i$$

$$\tilde{T}^k_m = r^k_{\cdot j} r^{\cdot i}_m T^j_i. \tag{8.72}$$

The transformation matrices are defined above as $r^k_{\cdot j} := \langle \tilde{g}^k, g_j \rangle$ and $r^{\cdot i}_m := \langle \tilde{g}_m, g^i \rangle$. This last line constitutes the change of basis formula for the second order tensor T of type $(1, 1)$. The extension of these principles to general tensors of type (r, s) is then clear, although it can be a notationally lengthy process.

It is now possible to interpret the particular form that these calculations take in this book. We restrict attention in this book to the choice of Euclidean space as the vector space U, $U = \mathbb{R}^3$. We also require orthonormal basis vectors. If $\{g_i\}_{i=1}^3$ is an orthonormal basis, the dual basis $\{\tilde{g}^j\}_{j=1}^3$ can be identified with $\{g_j\}_{j=1}^3$. The action of the dual basis on elements of $\mathbb{R}^3$ is then interpreted as the dot product

$$\langle \tilde{g}^j, g_i \rangle = g_j \cdot g_i = \delta^j_i := \delta_{ji}.$$

In this case the tensor transformation law in Equation 8.72 reduces to that introduced in Chapter 2. The general transformation law in Equation 8.72 holds in the event that the basis $\{g_i\}_{i=1}^3$ is not orthogonal nor made up of unit vectors, while the special case treated in Chapter 2 is applicable whenever we restrict attention to orthonormal basis vectors $\{g_i\}_{i=1}^3$.

S.3 Distributional and Weak Derivatives

Some care is required when we seek to interpret the evolutionary partial differential equations derived in this text such as those in Equations 7.21 and 7.23. In this brief section we review the definition of some of the spaces of classically differentiable functions, and we summarize the fundamental definitions of weakly differentiable functions. An introduction to these concepts can be found in [33], while more detailed accounts are given [1, 40], or [8]. Recall that if $f : \mathbb{R} \to \mathbb{R}$, we say that f is classically differentiable at $x \in \mathbb{R}$ if the limit

$$\frac{df}{dx}(x) := \lim_{\Delta x \to 0} \frac{f(x + \Delta x) - f(x)}{\Delta x}$$

exists. We define $C^k(\Omega)$ to be the collection of all k times continuously differentiable functions defined on the domain $\Omega \subseteq \mathbb{R}^n$. In other words, $C^k(\Omega)$ consists of functions that have a continuous derivative in Ω through order k. We define $C^k_b(\Omega)$ to be all those functions in $C^k(\Omega)$ whose derivative through k in Ω is also bounded. For example, if $f(x) = 1/x$ and $\Omega = (0, 1)$, then $f \in C^k(\Omega)$ for any $k \geq 0$. However, $f \notin C^k_b(\Omega)$ for any k. We define $C^\infty_0(\Omega)$ to be the collection of all smooth functions whose support is contained in a compact subset of Ω. The space $C^\infty_0(\Omega)$ is sometimes referred to as the space of test functions.

A careful inspection of the equations derived in Chapters 6 and 7 makes clear that some of the terms cannot be interpreted, strictly speaking, in terms of classical derivatives. For example, since the characteristic function $\chi_{[a,b]}$ is not continuous over $[a, b]$, the derivative $\frac{\partial^2 \chi_{[a,b]}}{\partial x_1^2}$ in Equation 7.23 does not exist in a classical sense. We must instead interpret these expressions in terms of distributional derivatives. We define a distribution T to be any continuous linear functional that acts on the test functions $C^\infty_0(\Omega)$. That is, an operator $T : C^\infty_0(\Omega) \to \mathbb{R}$ is a distribution if it is a linear and continuous operator. If T is a distribution, then we denote its action on any given $\phi \in C^\infty_0(\Omega)$ by $\langle T, \phi \rangle$. By definition, the distributional derivative ∂T of any distribution T is defined as the distribution that satisfies $\langle \partial T, \phi \rangle = -\left\langle T, \frac{d\phi}{dx} \right\rangle$ for all $\phi \in C^\infty_0(\Omega)$. With these definitions, it is not difficult to prove that the distributional derivative of the Heaviside function $H(x)$ is the Dirac delta $\delta(x)$ function. Moreover, the kth derivative of the delta function $\delta(x)$ is the distribution that maps $\phi \mapsto (-1)^k \phi^{(k)}(0)$ [40]. Together these two observations imply that terms like $\frac{\partial^2 \chi_{[a,b]}}{\partial x^2}$ that appear in the evolutionary equations for the piezoelectrically actuated composite beam can be interpreted as a linear combination of the second distributional derivative of translates of the Dirac delta function. This fact is used commonly in the engineering literature, see [42], where it is written that

$$\frac{\partial^2 \chi_{[a,b]}}{\partial x^2} = \frac{\partial^2}{\partial x^2}(H(x - a) - H(x - b)) = \frac{\partial \delta(x - a)}{\partial x} - \frac{\partial \delta(x - b)}{\partial x}.$$

We can thereby interpret Equations 7.21 and 7.23 rigorously in the sense of distributions.

Many of the distributions used in analysis are generated by functions. Any locally integrable function f defines a distribution T_f when we set

$$\langle T_f, \phi \rangle := \int_\Omega f(x)\phi(x)dx$$

for all $\phi \in C_0^\infty(\Omega)$. The function space $L^p(\Omega)$ contains (the equivalence classes of) functions whose p^{th} power is Lebesgue integrable. It is a Banach space with norm $\| u\|_{L^p} := \left(\int_\Omega |u|^p dx\right)^{1/p}$ for $1 \leq p < \infty$. When $p = \infty$, it is a Banach space with norm $\| u\|_{L^\infty} := \operatorname{esssup}_{x\in\Omega} |f(x)|$. By using the fact that every locally integrable function defines a distribution, we note that a function $u \in L^p(\Omega)$ is said to have a weak or distributional derivative $v := \frac{\partial u}{\partial x_k}$ if we have

$$\int_\Omega v(x)\phi(x)dx = -\int_\Omega u(x)\frac{\partial \phi}{\partial x_k}dx$$

for all test functions $\phi \in C_0^\infty(\Omega)$. In other words, the distributional derivative of a function satisfies the integration by parts formula with all smooth functions ϕ that vanish in a neighborhood of the boundary of Ω. The Sobolev spaces $W^{k,p}(\Omega)$ are constructed from collections of such weakly differentiable functions. We define the multi-integer $s = [s_1, s_2, \ldots, s_n]$ where each s_i is a non-negative integer and its norm $|s| = \sum s_i$. A mixed partial derivative is then written in the form $\frac{\partial^{|s|}u}{\partial x^s} := \frac{\partial^{|s|}u}{\partial x^{s_1}\ldots\partial x^{s_n}}$. The Sobolev space $W^{k,p}(\Omega)$ for $1 \leq p \leq \infty$ is the collection of functions $u \in L^p(\Omega)$ that have distributional derivatives $\frac{\partial^{|s|}u}{\partial x^s}$ that are contained in $L^p(\Omega)$ for multi-integers $|s| \leq k$. In other words, we define

$$W^{k,p}(\Omega) := \left\{ u \in L^p(\Omega) \,\middle|\, \| u \|^p_{W^{k,p}} := \sum_{|s|\leq k} \left\| \frac{\partial^{|s|}u}{\partial x^s} \right\|^p_{L^p} \leq \text{constant} < \infty \right\}.$$

The Sobolev spaces $W^{k,p}(\Omega)$ are Banach spaces with the definition of the norm $\| \cdot\|_{W^{k,p}}$ above. By convention, we denote the Hilbert space $H^k(\Omega) := W^{k,2}(\Omega)$, which is equipped with the inner product

$$(u, v)_{H^k} := \sum_{|s|\leq k} \int_\Omega \frac{\partial^{|s|}u}{\partial x^s}\frac{\partial^{|s|}v}{\partial x^s}dx.$$

The Sobolev spaces are used to make a precise statement of the evolutionary partial differential equations in this text. See [4] for more details on the role of weak formulations for piezoelectrically actuated thin plate and shell structures.

Bibliography

1 Robert A. Adams and John J.F. Fournier, *Sobolev Spaces*, Academic Press, Amsterdam, 2003.

2 U. Aridogan, I. Basdogan, A. Erturk, "Analytical Modeling and Experimental Validation of a Structurally Integrated Piezoelectric Energy Harvester on a Thin Plate," *Smart Materials and Structures*, Vol. 23, pp. 1–13, 2014.

3 U. Aridogan, I. Basdogan, A. Erturk, "Multiple Patch-Based Broadband Piezoelectric Energy Harvesting on Plate-Based Structures," *Journal of Intelligent Material Systems and Structures*, Vol. 25, No. 14, pp. 1664–1680, 2014.

4 H.T. Banks, R.C. Smith, and Y. Wang, *Smart Material Structures Modeling, Estimation, and Control*, John Wiley & Sons, Inc., Masson, Paris, 1996.

5 Etienne Balmes and Arnaud Deraemaeker, *Modeling Structures with Piezoelectric Materials: Theory and SDT Tutorial*, SDTools, Inc, Paris, France, 2014.

6 Eric Becker, Graham F. Carey, and J. Tinsley Oden, *Finite Elements: An Introduction, Volume 1*, Prentice-Hall, 1978.

7 Alain Bensoussan, Giuseppe Da Prato, Michel C. Delfour, and Sanjoy K. Mitter, *Representation and Control of Infinite Dimensional Systems: Volume I*, Birkhauser, 1992.

8 Susanne C. Brenner and L. Ridgway Scott, *The Mathematical Theory of Finite Element Methods,* Springer-Verlag, New York, 1994.

9 Francesco Bullo and Andrew D. Lewis, *Geometric Control of Mechanical Systems*, Springer, 2005.

10 Walter Cady, *Piezoelectricity: Volumes 1 and 2*, Dover, 1964.

11 Roy R. Craig and Andrew J. Kurdila, *Fundamentals of Structural Dynamics*, John Wiley & Sons, Inc., 2006.

12 Stephen H. Crandall, Dean C. Karnopp, Edward F. Kurtz, and David C. Pridmore-Brown, *Dynamics of Mechanical and Electromechanical Systems*, McGraw-Hill, Inc., 1968.

13 A. Cemal Eringen and Gerard A Maugin, *Electrodynamics of Continua I*, Springer–Verlag, Inc., New York, 1990.

Vibrations of Linear Piezostructures, First Edition. Andrew J. Kurdila and Pablo A. Tarazaga.

This Work is a co-publication between John Wiley & Sons Ltd and ASME Press.

14 Alper Erturk and Dan Inman, *Piezoelectric Energy Harvesting*, John Wiley & Sons, 2011.

15 Y.C. Fung, *A First Course in Continuum Mechanics*, Prentice Hall, Inc., 1977.

16 David J. Griffiths, *Introduction to Electrodynamics*, Pearson, Boston, 2013.

17 Herbert Goldstein, *Classical Mechanics*, Addison-Wesley Publishing Company, Inc., 1980.

18 Jill Guyonnet, *Ferroelectric Domain Walls*, Springer, 2014.

19 Thomas J.R. Hughes, *The Finite Element Method: Linear Static and Dynamic Finite Element Analysis*, Dover, 2000.

20 T. Ikeda, *Fundamentals of Piezoelectricity*, Oxford University Press, 1996.

21 Daniel J. Inman, *Engineering Vibration*, Pearson Education, Inc., New Jersey, 2014.

22 B. Jaffe, W.R. Cook, and H. Jaffe, *Piezoelectric Ceramics*, New York, Academic Press, 1971.

23 Maureen M. Julian, *Foundations of Crystallography*, CRC Press, 2008.

24 Erwin Kreyszig, *Introductory Functional Analysis with Applications*, John Wiley & Sons, Inc., 1978.

25 Andrew J. Kurdila and Michael Zabarankin, *Convex Functional Analysis*, Birkhauser Verlag, Boston, 2005.

26 Donald J. Leo, *Smart Material Systems*, John Wiley & Sons, Inc., New York, 2007.

27 Vijaya Venkata Malladi, *A Study of Piezoceramics Induced Traveling Waves through Simulations and Experiments, PhD Dissertation*, Virginia Tech, 2016.

28 Cecil Malgrange, Christian Ricolleau, and Michel Schlenker, *Symmetry and Physical Properties of Materials*, Springer, 2014.

29 Leonard Meirovitch, *Analytical Methods in Vibrations*, MacMillan, New York, 1975.

30 Leonard Meirovitch, *Fundamentals of Vibrations*, Waveland Press, 2012.

31 Walter J. Merz, *Domain Formation and Domain Wall Motion in Ferroelectric $BaTiO_3$ Single Crystals, em Physical Review*, Volume 95, Number 3, August 1, 1954, pp. 690–698.

32 Arch W. Naylor and George Sell, *Linear Operator Theory in Engineering and Science*, Springer-Verlag, New York, 1982.

33 J. Tinsley Oden, *Applied Functional Analysis*, Prentice-Hall, 1979.

34 J. Tinsley Oden, Graham F. Carey, and Eric B. Becker, *The Finite Element Method: An Introduction, Volume 1*, Prentice Hall, Inc., 1983.

35 Katsuhiko Ogata, *System Dynamics*, Pearson, 2014.

36 PCB Piezotronics, `http://www.pcb.com/`.

37 PI Ceramics, `https://www.piceramic.com`.

38 A. Preumont, *Mechatronics: Dynamics of Electromechanical and Piezoelectrical Systems*, Springer, 2006.

39 J.N. Reddy, *An Introduction to the Finite Element Method,* McGraw-Hill, Inc., 2006.

40 Michael Renardy and Robert C. Rogers, *An Introduction to Partial Differential Equations*, Springer-Verlag, New York, 1993.

41 Ralph E. Showalter, *Hilbert Space Methods for Partial Differential Equations*, Dover, 2010.

42 Samuel C. Stanton, Alper Erturk, Brian P. Mann, and Daniel J. Inman, "Nonlinear Piezoelectricity in Electroelastic Energy Harvesters: Modeling and Experimental Identification," *Journal of Applied Physics*, Vol. 108, 2010.

43 Jan Tichy, Jiri Erhart, Erwin Kittinger, and Jana Privratska, *Fundamentals of Piezoelectric Sensorics*, Springer–Verlag, Inc., Berline, 2010.

44 H.F. Tiersten, *Linear Piezoelectric Plate Vibrations,* Springer Science+Business Media, 1969.

45 Joseph F. Vignola, John A. Judge, and Andy Kurdila, *Shaping of a System's Frequency Response Using an Array of Subordinate Oscillators, J. Acoust. Soc. Am*, Vol. 1, pp 129–139, July, 2009.

46 J. Wloka, *Partial Differential Equations*, Cambridge University Press, 1987.

47 Jiashi Yang, *An Introduction to the Theory of Piezoelectricity*, Springer, 2005.

48 Jiashi Yang, *Analysis of Piezoelectric Devices,* World Scientific Publishing, 2006.

49 Yi–Yuan Yu, *Vibrations of Elastic Plates: Linear and Nonlinear Dynamical Modeling of Sandwiches, Laminated Composites, and Piezoelectric Layers*, Springer, New York, 1996.

50 Andrew Zangwill, *Modern Electrodynamics*, Cambridge University Press, Cambridge, 2013.

51 O.C. Zienkiewicz, R.L. Taylor, and D.D. Fox, *The Finite Element Method for Solid and Structural Mechanics,* Elsevier, 2000.

52 Ralph Abraham and Jerrold E. Marsden, *Foundations of Mechanics*, AMS Chelsea Publishing, 2008.

53 Walter Guyton Cady, *Piezoelectricity: An Introduction to the Theory and Applications of Electro-mechanical Phenomena in Crystals*, Dover Publications, Inc., 1964.

Index

This Work is a co-publication between John Wiley & Sons Ltd and ASME Press.